# *The* Water-Smart Garden

# *The* Water-Smart Garden

TECHNIQUES AND STRATEGIES
FOR CONSERVING, CAPTURING, AND
EFFICIENTLY USING WATER IN TODAY'S
CLIMATE . . . AND TOMORROW'S

**Noelle Johnson**

*For my dear husband who has been my constant encourager along my garden journey. Without you, this book wouldn't exist.*

**Quarto.com**


First Published in 2025 by Cool Springs Press,
an imprint of The Quarto Group,
100 Cummings Center, Suite 265-D,
Beverly, MA 01915, USA.
T (978) 282-9590 F (978) 283-2742



Cool Springs Press titles are also available at discount for retail, wholesale, promotional, and bulk purchase. For details, contact the Special Sales Manager by email at specialsales@quarto.com or by mail at The Quarto Group, Attn: Special Sales Manager, 100 Cummings Center, Suite 265-D, Beverly, MA 01915, USA.

29 28 27 26 25 1 2 3 4 5

ISBN: 978-0-7603-8824-2

Digital edition published in 2025
eISBN: 978-0-7603-8825-9

Library of Congress Cataloging-in-Publication Data is available.

Design and Page Layout: Laura Shaw Design
Photography: See credits on page 182
Illustration: Holly Neel

Printed in Malaysia

**Cover (clockwise, from top left):** A rain barrel collects water from a roof in Minneapolis, Minnesota; a Texas courtyard garden with low-water flowering plants; a xeriscape garden with lance leaf coreopsis (*Coreopsis lanceolata*) along with Rocky Mountain penstemon (*Penstemon strictus*) at Denver Botanic Gardens Chatfield Farms; and a vegetable garden in Washington state with drip irrigation. **Page 2:** A dry creek bed directs rainwater away from a house toward plants in a Texas garden. **Page 5:** Mexican bush sage (*Salvia leucantha*) stands out amid Mexican feather grass (*Nassella tenuissima*) in a drought-tolerant garden. **Page 6:** Purple coneflower (*Echinacea purpurea*)

# Contents

# Welcome to Gardening with Less Water

**Fresh water is a precious resource,** with only a small fraction available for our use. With climate change, temperatures are heating up, and rainfall amounts are decreasing in many regions. As a result, we must increase the supplemental water to the plants in our gardens to ensure their survival. Having to water plants more puts additional strain on water supplies. Unsurprisingly, this can be stressful for you and your plants too—especially if your plants require frequent watering. Whether you live in a dry climate, have experienced long-term or brief periods of drought, or want to ensure that you use water as efficiently as possible in your garden—this book is for you!

In drier regions, you don't need to look far to see rivers slowing to a small trickle, drying up, or lake levels falling dramatically. Groundwater supplies throughout many regions also are decreasing quickly due to drought and water overuse. With less rain, water is being pumped from groundwater supplies, often faster than it is replaced, which is occurring in both dry climate regions and those that experience higher rainfall amounts. In many coastal regions, salt water is permeating fresh-water aquifers due to over-pumping. In short, we are using more water than is being replenished, which isn't sustainable.

The incidence of drought is increasing worldwide and isn't just restricted to dry climate regions. Less-than-average rainfall, shrinking water supplies, and irresponsible water practices lead to higher water bills. In many areas, decreasing water supplies have led to watering restrictions regarding how often and how much water a household can use.

The Colorado River flows out from the Hoover Dam in the Southwest region of the United States.

Raindrops on a southwestern prickly poppy (*Argemone pleiacantha*) after months of drought in a California garden.

The author explores the beauty of drought-tolerant plants in a California garden close to where she grew up.

My water journey began in California as a child, and I vividly remember episodes of drought where we had to be mindful of our water usage. I now call the deserts of Arizona my home, where less rain falls, and water is even more of a precious resource. I have a beautiful garden that doesn't use a lot of water, and I'm excited to share how you can have a water-smart garden, whether you live in a region where water is scarce or where rainfall supplies most of your water needs. Our overuse of water invites us to reexamine our relationship with water and not take it for granted.

The strategies within these pages consist of an overall approach for whatever region you call home. As you begin (or continue) your water-saving journey, we want to focus on outdoor water use, where a surprisingly high percentage of water use occurs. Household water use for the lawn and plants around your home is higher than you think—ranging from 30 up to 70 percent. That is a lot of water! But that means we can substantially lower our water use by focusing on the areas outside your house. We can implement water efficiency in how we design our gardens, choose the right plants, capture and direct rainwater, and water our plants the "right" way. In other words, we need to reimagine what a garden should be—instead of one that requires a lot of water, how about one with beauty but requires less water? Incorporating strategies such as choosing the right plants, implementing water-efficient irrigation practices, and making the most of rain, along with other methods within this book, will enable you to use water wisely while enjoying a beautiful outdoor space around your home.

If your vision of a waterwise garden is one devoid of plants or solely comprised of cacti and rocks, I'm here to show you how you can have a beautiful garden filled with a variety of plants while lowering your water use. To do this, we need to create resilient gardens that make the most of what water they receive while identifying ways we are wasting water. One crucial way to achieve this is to reexamine what plants we choose, opting for those that look great but don't require a lot of water. I am constantly amazed at the beauty of low-water plants! You don't need to sacrifice beauty for water efficiency—you can have both.

Imagine a garden that thrives with less water, adds beauty, and thrives in your outdoor space. We will explore how to create and maintain a water-efficient outdoor space. You will gain valuable information and inspiration with the pages of this book, whether you live in a dry climate or want to save water with your gardening practices. To do this, we will examine the common mistakes that lead to higher water use, including overwatering plants, high-water-use plants, and inefficient watering methods. Later, we will explore ways to capture and direct rainwater and explore different ways of watering plants to help you select the best watering choice.

Soil type and how we maintain plants also play a part in lessening a plant's water needs. Changes

A waterwise Texas garden that thrives primarily on rainfall.

to your garden may be needed to conserve water, including removing your lawn and replacing it with attractive plants requiring less water and maintenance. Gardening can be fun, even in lower water conditions, and I'm excited to share water-saving tips for your container and vegetable gardens. Finally, at the end of this book, you'll find garden projects to implement to help save water.

Let's begin your water-smart garden journey!

CHAPTER 1

# Identifying Low-Water Conditions Where You Live

**When we think about water availability,** we tend to focus on drier regions with annual rainfall of 20 inches (51 cm) or less. Or areas where long-term drought has shrunk water supplies. However, it may surprise you that shrinking water supplies also are occurring in more temperate regions and they are having to deal with water availability issues. Often, this is due to using more water than is replenished from rain. In many areas, groundwater is being pumped out of the ground faster than it is replaced, and increasing periods of drought only exacerbate this practice.

Drought-resistant Blue Bells™ emu (*Eremophila hygrophana*), pink muhly grass (*Muhlenbergia capillaris*), and angelita daisy (*Tetraneuris acaulis*) in the author's back garden.

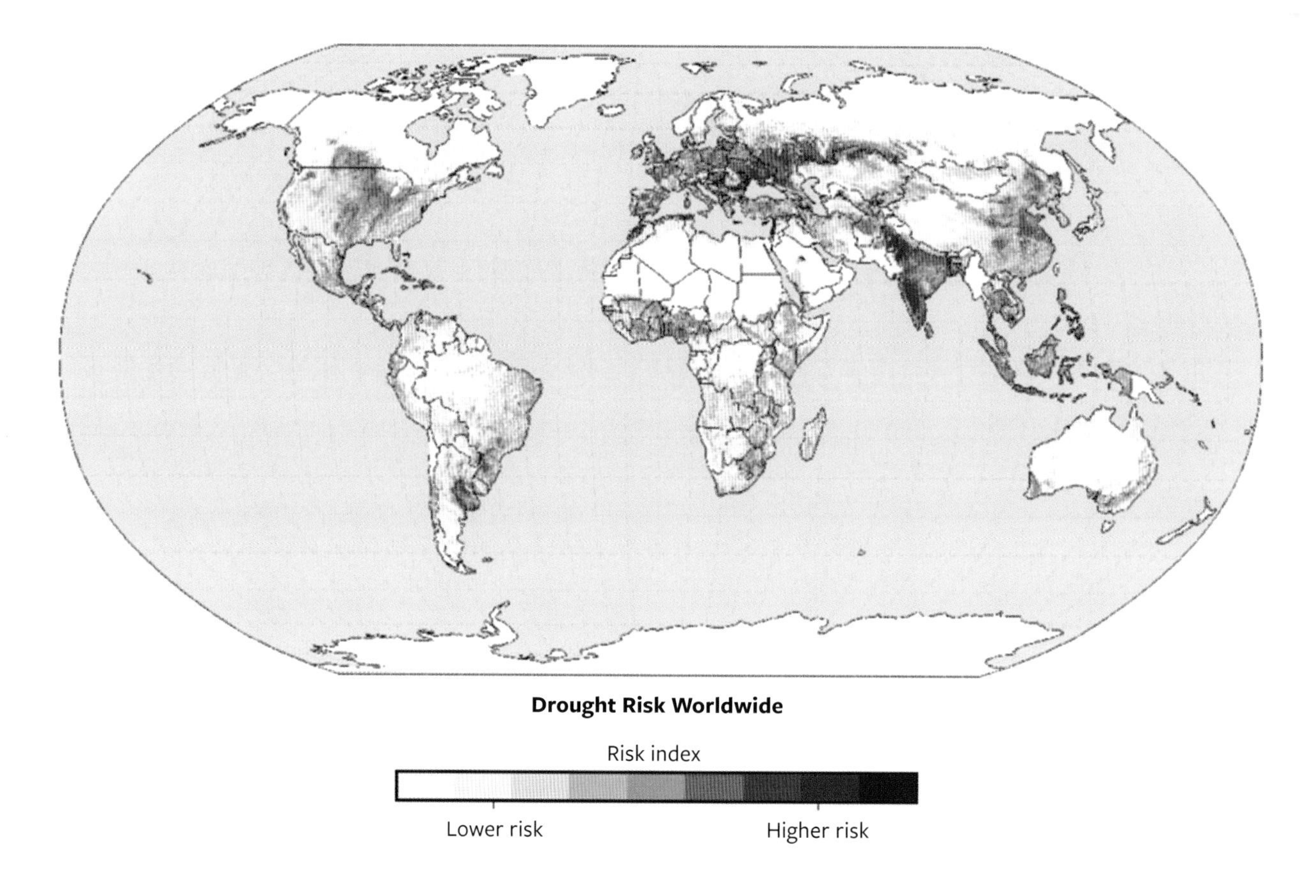

**Drought Risk Worldwide**

Risk index

Lower risk Higher risk

Source: Vogt, Jürgen & Naumann, Gustavo & Masante, Dario & Spinoni, Jonathan & Cammalleri, Carmelo & Erian, Wadid & Pischke, Frederik & Pulwarty, Roger & Barbosa, Paulo. (2018). Drought Risk Assessment and Management. A Conceptual Framework. 10.2760/919458.

## What Is Drought?

In the past, throughout many developed regions of the world, there hasn't been much thought about the availability of water when people turn on water to irrigate their plants. However, in recent years, shrinking water supplies and a warming climate have brought the word *drought* into the mainstream. For people who live in arid or semi-arid regions, dry periods aren't uncommon, yet drought is becoming more prevalent in recent years. In wetter areas, drought is happening more regularly due to our warming climate. Whether your region is dealing with a decades-long drought or shorter-term drought conditions, it's a factor that we need to incorporate into our gardening practices to ensure that we use water efficiently.

But what is drought exactly? Drought occurs when a region experiences less rainfall than usual over a period of time. As a result, river flows lessen, lake levels dip, and plants and animals in nature suffer from a lack of water, which creates water stress. While arid and semi-arid regions tend to grab the headlines with periods of drought, they aren't the only places where drought happens. Temperate, tropical, and semi-tropical areas where rainfall is ordinarily plentiful can and do experience drought.

## How Does Drought Affect You?

When our water needs outweigh what rainfall supplies, we get into trouble. In the garden, that can be overwhelming and stressful when your plants become more reliant on supplemental water because of a lack of rain. Whether your plants survive on rainfall alone or need water regularly, periods of

A waterwise garden in Bellevue, Washington, planted with nepeta (*Nepeta* spp.) and the airy flowers of gaura (*Oenothera lindheimeri*).

drought will affect them. If you have experienced less rainfall than usual, you may have noticed its effects on your plants—wilting and yellowing leaves or even the death of the plant. Drought stress can happen even if you provide irrigation at your regular frequency because rainfall can play a large part in a plant's total water allowance. Native plants that grow well under average rain amounts aren't immune to drought stress either. Smaller leaf size, fewer leaves, wrinkled succulent leaves, and yellowing foliage are signs native plants will show when not receiving enough water from rain.

Drought directly affects you when your garden doesn't look its best due to struggling plants, plant loss, and a higher water bill from the extra water you provide for plants. Understandably, this can make

A flowering groundcover showing signs of drought stress with browning foliage.

Annual Precipitation Minus Evaporation

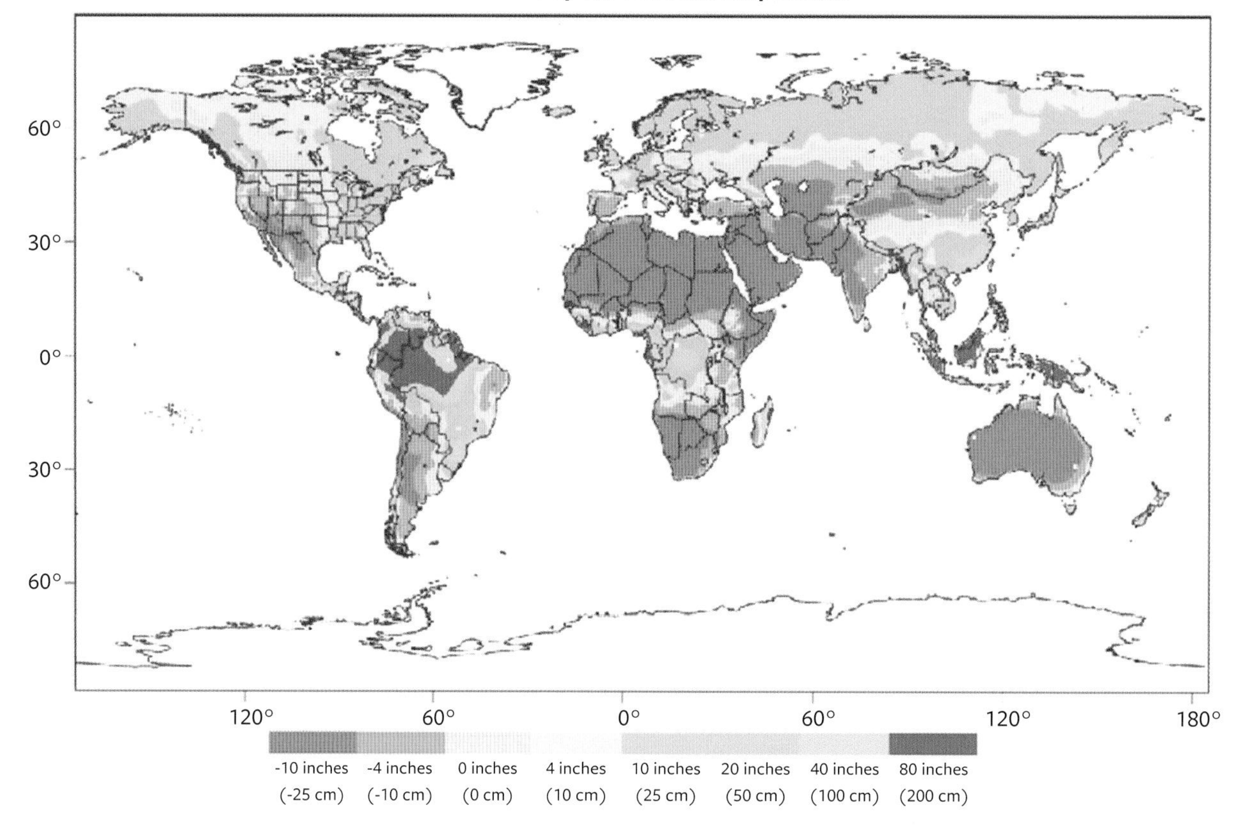


Color-coded map based on annual rainfall a region receives minus the amount of water that evaporates into the air. Look for plants native to regions with similar rainfall and evaporation rates as where you live. Source: Givnish, Thomas. (2002). Adaptive Significance of Evergreen vs. Deciduous Leaves: Solving the Triple Paradox. Silva Fennica. 36. 10.14214/sf.535.

you feel stressed and overwhelmed. If this sounds like you, you may be ready to switch out thirsty plants for attractive, low-water alternatives while exploring other ways to reduce water usage in the garden.

Because most water use occurs outdoors, creating a water-efficient garden can significantly affect the amount of water you use and result in a lower water bill. Thankfully, the strategies to conserve water around your home don't take away from the prospect of an attractive outdoor space—waterwise landscapes can be beautiful! Moving forward, we must create drought-resilient gardens filled with low-water plants uniquely suited to our region. The bonus is that whether you are experiencing drought conditions or not, your garden will be ready to meet the challenges and use water efficiently.

To begin, let's identify if your region is experiencing drought. Drought maps are available online (see links provided in the Resources section). Drought can be long-lasting or come on quickly within a few months. Therefore, it's a good idea to check regional drought maps periodically. To further understand where you live and the effects on your garden, it's essential to know the average amount of rainfall you receive.

Next, identify what months you typically receive rain. Not all gardens get rain every month. Your rainfall may occur seasonally, such as in winter and spring, while summers can be dry. Alternatively, some regions see rain in the summer months. Often, there are seasonal rain patterns where rainfall is more prevalent than at other times—for example, in my desert garden, the majority of rain falls in winter and summer. When you receive rain will affect what plants you select and how you configure your irrigation system seasonally; see chapters 2 and 6.

Are there areas around the landscape outside your home that are perpetually dry? Sites at a higher level than others in the garden tend to be drier than areas at the base of a slope or gentle swale. If parts of your garden are in full sun throughout most of the day, they will be drier than those in the shade. These drier zones of your landscape are where drought or water stress will likely be seen first and need intervention most.

The road to creating a water-smart garden is to know if you are in a drought, what your average rainfall is, and when it typically occurs. Finally, identify where the driest areas around your home are. Once we know these factors, we can implement strategies to reduce our water use in the garden. Even if you don't live in a dry region or one that is affected by drought, understanding the climate where you live will allow you to make the best decisions as to plant choice, watering practices, and maintenance.

A combination of native and adapted plants thrive in this Perth, Australia, garden, including Frosty Kiss™ Orange Flame gazania (*Gazania rigens* Frosty Kiss™ Orange Flame), which is native to South Africa along with Australian natives such as willow myrtle (*Agonis flexuosa*) and cotton heads (*Conostylis candicans*).

CHAPTER 2

# Plant Choice and How It Influences Water Use

## Thirsty Plants vs. Low-Water Plants

All plants have varying water needs, depending on the type and where each one comes from. Usually, the region a plant is native to determines how much water they require to survive. That means that a plant native to a drier climate won't need a lot of water to grow, while a plant from a region with ample rainfall will require more water. A plant's water needs depend on how they successfully evolved with the climatic and soil conditions they have had to contend with. If you try to grow a plant from a temperate or tropical region in drier conditions, it will require more water to survive because rainfall won't be enough. In other words, the more we use plants that can survive on water coming from rain, the less we need to use supplemental water resources. If you figure in rainwater harvesting, you can increase the variety of plants you can grow without using secondary water resources.

Low-water plants aren't boring. At the Denver Botanic Gardens Chatfield Farms, the colorful flowers of coreopsis (*Coreopsis lanceolata*), foxtail lily (*Eremurus robustus*), Rocky Mountain penstemon (*Penstemon strictus*), and globe mallow (*Sphaeralcea ambigua*) add welcome beauty to this drought-tolerant garden.

## WHEN NATIVE PLANTS NEED AN EXTRA DRINK

Native plants and their cultivars are an excellent choice for a drought-resilient garden as they can generally survive on average rainfall amounts where you live. However, they have small root systems when first planted and will need to be watered by you until they grow a more extensive root system. Sufficient root growth can take about a year for shrubs and groundcovers, while trees may need up to three years of supplemental water to become established. It's important to note that in drought, established native plants may require additional water periodically to replace the lack of rain.

Red emu bush (*Eremophila maculata*) is a drought-tolerant shrub native to Australia that must have well-drained soil. This one died due to being overwatered due to its close location to the lawn.

Plants with lower water needs can be as lush and beautiful as thirsty ones. While higher water-use plants can struggle under drier climatic conditions, low-water plants can have difficulty growing in areas that receive a lot of rain, where soil seldom dries out. Plants with lower water needs generally require well-drained soil to do their best because they have higher oxygen requirements from the soil. Oxygen is present within the tiny spaces between individual soil particles. When we water plants, those little spaces fill with water, temporarily forcing oxygen out. When soil stays wet for long periods or seldom dries out, there is less oxygen present for plants. Plants from regions with higher rainfall are generally better adapted to consistently moist soils, while low-water plants are not.

### How Plants Use Water

Water is a big deal for plants, as they are made of up to 95 percent water. They use water to grow, move nutrients, carry on photosynthesis, and provide structural strength to plant cells. Water moves through the plant, often in response to the sun, and is released into the atmosphere by the leaves through tiny openings called stomata in a process called transpiration, which is how a plant moves water and releases oxygen made during photosynthesis. The sun is the primary driver in this process—the more sun, coupled with heat, the more water is lost through the foliage of plants. This process helps to keep a plant from overheating (like people do when they sweat) and allows photosynthesis to occur. Water needs to be replaced in plants as it's lost to the atmosphere. When a plant doesn't receive enough water, we see wilting due to loss of cell strength, browning foliage, yellowing due to nutrient loss, and generally less growth. Using plants that require less water is vital to a thriving garden in low-water conditions.

## What Types of Plants Are Best for Drought-Resilient Gardens?

Plants adapted to lower water conditions also have specific characteristics and strategies to help them conserve water that their thirsty cousins do not. When shopping for plants, look for certain characteristics that may indicate adaptations to low-water conditions. Conversely, some attributes point to a plant needing regular irrigation. Of course, the best way to ensure you select the right lower water-use plant for your garden is to research plants before buying.

### Drought-Tolerant Plant Characteristics

What makes a plant waterwise is its ability to conserve water, and the most common way plants do this is to limit water loss from the leaves. Plants lose water through their stomata, which are tiny pores on the leaf surface. There are several leaf characteristics that plants employ to conserve water. Plants with large leaves have a greater surface area and tend to lose more water. Plants that are in the lower water-use group often have a smaller-sized leaf, meaning less surface area for water to leave the plant. During drought, low-water plants may form smaller leaves than usual to limit water loss even further. Another adaptation is to see plants having foliage with a grayish hue that gives a blue-gray or green-gray color from tiny hairs covering the leaf. These hairs help reflect the sun's rays and prevent too much water from leaving the stomata. Many low-water-use plants have a thick waxy coating on the leaves that helps to lessen water loss through the leaves.

You may have noticed that some desert regions have shrubs with green stems, like chuparosa (*Justicia californica*), and palo verde trees (*Parkinsonia* spp.) with green branches and trunks. These green plant parts offer a valuable adaptation during extreme drought conditions as the plants will shed their leaves to conserve water, and the process of

The fine-textured foliage of moss verbena (*Glandularia aristigera*) and lavender cotton (*Santolina chamaecyparissus*) are characteristic of many types of low-water-use plants, which reduce water loss from leaves.

One characteristic that many drought-tolerant plants share is tiny hairs covering the leaves, such as this little leaf cordia (*Cordia parvifolia*), which helps limit water loss.

Tall verbena (*Verbena bonariensis*) highlights the blue-gray leaves of succulent silver spurge (*Euphorbia rigida*) and yucca in a Seattle, Washington, garden.

photosynthesis will carry on through their other green plant parts.

Cacti and other succulents store water inside for use when water supplies are scarce. They have a thick, waxy coating that helps to prevent water loss. Water loss (transpiration) from this group of plants occurs through the surface of the leaves, pads, and stems. But how do they prevent water from being lost through their stomata during the day? Instead of opening their stomata in the daytime, they do it at night, in a process called CAM photosynthesis, which means that less water is lost because there is no sun to generate high levels of transpiration.

## What Does "Drought-Tolerant" Really Mean When It Comes to Plants?

If you are looking into using less water in the garden, you've probably heard the terms *drought- tolerant*, *drought-resistant*, *low-water*, or *waterwise* used to describe plants or gardens. But what do these terms mean? Does that mean plants with such a description don't need any supplemental water? The fact is that there is a lot of confusion around these terms and what they mean. Often, this is due to people using them interchangeably, and the definitions of each are fluid and changing. To further confuse things, these terms mean different things regarding water tolerance depending on the region and its climate. A plant may be considered drought-tolerant in one climate, while it would struggle to survive in a drier area without extra water.

We can agree that all these terms describe plants that perform well in drier conditions—in other words, they are lower water-use plants. However, that doesn't mean they don't require supplemental water during times of drought or even regularly in an arid region such as the desert. In a severe drought with little to no rain over several months, you may need to provide supplemental water to so-called drought-tolerant (drought-resistant, waterwise) plants if they begin to show signs of stress. For many desert regions, many drought or desert-adapted plants require regular watering throughout their lifespan.

It's helpful to think of plants identified by low-water terms such as *drought-tolerant* as plants that can tolerate less than average rainfall amounts or those that may need minimal supplemental irrigation. Drought-tolerant doesn't mean the plant will be fine without any additional water, no matter the level of drought conditions. Your best strategy is to look to local watering guidelines, monitor your plants, and provide water to them if they show signs of drought stress.

The green stems of chuparosa (*Justicia californica*) carry on photosynthesis in drought conditions when the plant sheds its leaves to save water.

## Why Natives Are Best Suited for Water-Smart Gardens

Thoughtful plant choice is essential when it comes to saving water around the landscape. For gardens that require frequent watering, a high water bill and struggling plants due to water restrictions are possibilities. Thankfully, it is possible to have a beautiful and thriving garden filled with plants that need less water. The first group of plants to look at when selecting plants for your home should be those native to your region. In many cases, they are the most water-efficient choice because they are adapted to growing, flowering, and thriving on natural rainfall amounts where you live, and only need supplemental water during dry periods. Another bonus is that they tend to be more resistant to insect damage and disease and don't need you to fertilize them. I am a strong proponent of plants that look great with little fuss or maintenance required, and native plants (or cultivars of natives) are a great place to start.

## Looking Beyond Natives to Adapted Plants

If you are looking to expand beyond the native plant palette in your garden while keeping an eye on low-water usage, look for plants native to regions with similar climatic conditions. Plants from areas with temperature, humidity levels, and rainfall amounts that resemble your climate can be an additional waterwise option for your garden. Adapted plants can add beauty around your home and attract pollinators. A word of caution: it is important to determine whether a particular plant can be invasive where you live. For example, lantana is a popular groundcover and shrub that performs beautifully in drier regions, but it can be invasive if grown in warm, humid climates. A plant's invasiveness often depends on what environment it grows in. Your local nursery professional or some online sleuthing can help provide these answers.

A little research using the map on page 26 can help you determine what areas around the globe closely match yours in regard to temperature, humidity, and rainfall. Once you've identified regions with a similar climate to yours, do an online search for plants native to that region to determine which ones may thrive in your region. In my desert garden, there are plants native to the American Southwest alongside aloes from Africa and colorful Australian shrubs—all of which are lower water use.

### HOW TO TELL IF YOUR PLANTS NEED WATER

Look for signs of drought stress in your plants to determine if they need more water than they receive from rainfall (or need more frequent supplemental irrigation). Signs to look for include:

» Smaller leaves than usual
» Yellowing or browning of foliage
» Unusual leaf drop
» Absence of flowers
» Wrinkling of cacti and succulents

**top** A colorful landscape filled with drought-tolerant plants native to the Southwestern region of the United States that include desert marigold (*Baileya multiradiata*), brittlebush (*Encelia farinosa*), Santa Rita prickly pear cactus (*Opuntia santa-rita*), and Parry's penstemon (*Penstemon parryi*).

**bottom** A collection of native and nonnative plants from Brazil, the American Southwest, and Southern Europe, which are well-adapted to the Mediterranean climate of this Santa Barbara, California, garden. Plants include agave (*Agave* spp.), torch aloe (*Aloe arborescens*), bougainvillea (*Bougainvillea* spp.), cane cholla cactus (*Cylindropuntia spinosior*), alyssum (*Lobularia maritima*), and blue chalksticks (*Senecio serpens*).

**World Climate Zones Based on Temperature and Precipitation**

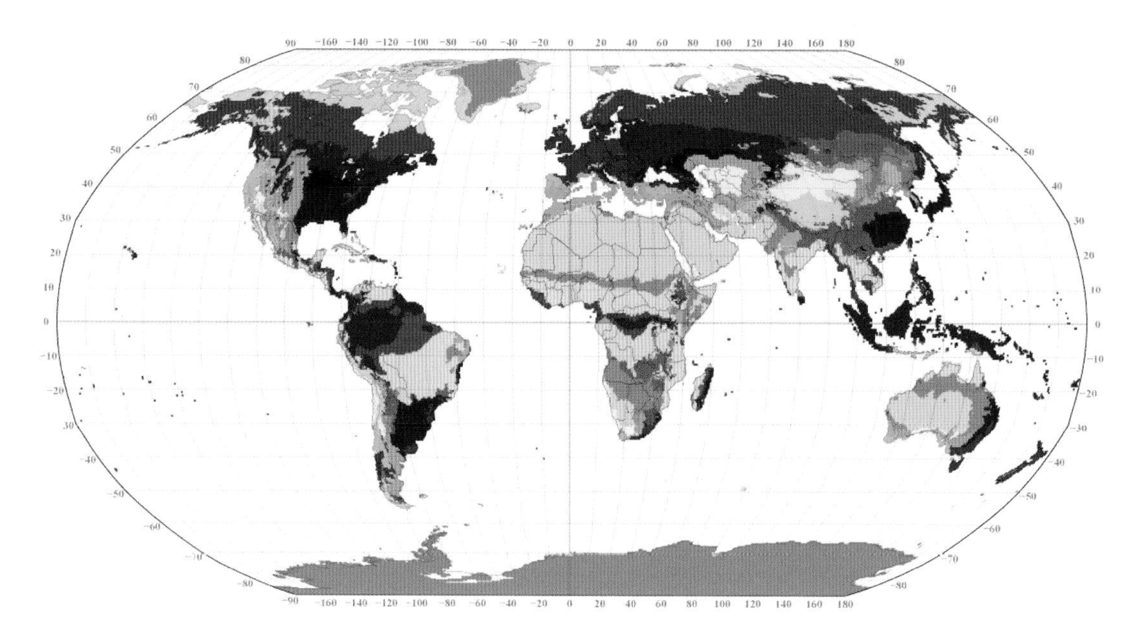

Climate zones incorporate precipitation and temperature, along with other variables. Look for plants from regions indicated by the same color as yours. Source: National Oceanic and Atmospheric Administration

## Avoid Using Plants from Higher Rainfall Regions

We get into trouble with higher water usage when we incorporate plants with much greater water needs into drier climates, such as those from wet regions of the world, which include both tropical and temperate climates. They need a lot of water to grow well and remain healthy. In lower water situations, they will struggle. Many plants from these wetter regions are beautiful, but there are equally attractive plants that require less water. Our goal is to buy plants only after considering how much water they need.

If you live in a dry region such as the desert or suffer from drought, many plants in your garden may struggle to survive on rainfall alone. In that case, it's essential to manage water resources in your garden efficiently with irrigation choices, appropriate water scheduling, and employing water-harvesting strategies, see chapters 5 and 6.

# Favorite Low-Water Plants

Many plants are suitable for low-water conditions. Now, let's be clear—a plant that can tolerate and even thrive in low-water conditions doesn't mean that they are drab and unattractive. In fact, the opposite is often true! Drought-tolerant plants have many desirable traits, including colorful flowers, attractive foliage, and pleasing shapes—all of which can enable you to create a beautiful water-wise garden.

I am sharing my favorite waterwise plants and suggestions from gardening experts in other regions to get started. Look at the regional key (page 27) to see which region most closely matches yours to determine whether a particular plant on the list will be suitable for your garden. If you don't live in these regions, find the one that most closely resembles the climate where you live and use that specific plant list. To determine what plants will perform best in your garden, I have listed cold

## Climate Regions

|  |  |
|---|---|
| COASTAL | Perpetually wet winters with mild to cold temperatures and warm, dry summers are characteristic of this region. |
| SEMI-ARID | Regions with drier conditions. Most rainfall occurs in winter, and summers are hot and dry, and include Mediterranean climates. Precipitation ranges from 10 to 20 inches (25–51 cm) annually. |
| DESERT | Dry climates that experience hot summers where evaporation rates exceed the amount of rainfall—the average amount of rain per year averages 12 inches (30 cm) and under. |
| MOIST SUBTROPICAL | Regions where summer temperatures range from warm to hot coupled with humid conditions. Winter temperatures range from cold to mild. Rainfall occurs steadily throughout the year. |
| MOIST CONTINENTAL | Areas with distinct seasonal temperature variations. Rain occurs throughout the year. Summers are cool to warm with cold to severely cold winters. |
| HIGHLANDS | Mountainous regions with cold winters and relatively mild summers. Temperature and humidity decrease while precipitation increases the higher the altitude. |
| TROPICAL | Hot, humid, and rainy weather are hallmarks of these regions that are usually found close to the equator. There may be seasonal dry periods. Freezing temperatures are rare. |

tolerances for plants grown outdoors year-round. Optionally, many shrubs, groundcovers, perennials, and succulents can be grown outdoors in containers through the warmer months and brought indoors in winter if you live in a colder region.

Many plants are available in related varieties, which offer additional flower or foliage color, drought hardiness, and cold tolerance. While this list is by no means exhaustive, it does serve as a good starting point when used in conjunction with regional online plant guides and your local nursery professional, who can also let you know of the availability of plants in your area or provide additional or alternative suggestions.

For the following plant lists, I am using regions that include generally shared characteristics primarily based on precipitation, humidity, and temperature. However, there are notable differences within each climate region. These lists are just a small sampling of plants to consider, and not all will be suitable for your region. For additional plant options, visit the Resource section. To ensure that a particular listed plant will do well in your garden, consult with your local nursery professional.

Desert willow
(*Chilopsis linearis*)

Blue wild indigo
(*Baptisia australis*)

'Tangerine Beauty' crossvine (*Bignc*
*capreolata* 'Tangerine Beauty')

California lilac
(*Ceoanthus* spp.)

'Moonshine' yarrow
(*Achillea* × '*Moonshine*')

Silver spurge
(*Euphorbia rigida*)

Purple coneflower
(*Echinacea purpurea*)

Red yucca
(*Hesperaloe parviflora*)

'Green Cloud' Texas ranger
(*Leucophyllum frutescens* 'Green Cloud')

Prairie feather
(*Liatris spicata*)

Penstemon
(*Penstemon* spp.)

Upright prairie coneflower
(*Ratibida columnifera*)

'Lemon Coral' stonecrop
(*Sedum mexicanum* 'Lemon Coral')

Chaste tree
(*Vitex agnus-castus*)

Pink muhly grass
(*Muhlenbergia capillaris*)

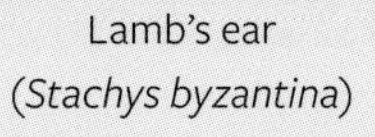

Lamb's ear
(*Stachys byzantina*)

Yellow bells
(*Tecoma stans*)

Bush germander
(*Teucrium fruticans* 'Azureum')

'Autumn Joy' sedum
(*Sedum spectabile* 'Autumn Joy')

**Let's start with trees,** which are our gardens' visual anchor, providing welcome beauty and shade. They also add value to your property, as most homeowners prefer homes with trees. The shade from trees also can reduce the water requirements of plants by providing protection from the drying effects of wind and the sun. Trees that provide dappled shade, rather than deep shade, allow you to grow a broader range of plants by allowing some sun to reach plants below.

## Trees

| NAME | REGION | MATURE SIZE | COLD HARDINESS | NOTES |
|---|---|---|---|---|
| **Bigtooth Maple** (*Acer grandidentatum*) | Moist Continental<br>Highlands | 30–40 × 20–30 ft (9.1–12.2 × 6.1–9.1 m) | -30°F (-34°C) | Deciduous, multi or single-trunk tree with lush-green, deeply lobed leaves. Vibrant orange foliage appears in fall. |
| **Mulga** (*Acacia aneura*) | Semi-Arid<br>Desert | 18 × 16 ft (5.5 × 4.9 m) | 15°F (-9°C) | Small tree with gray, narrow leaves. The foliage is evergreen and has a pyramidal or rounded shape. It has a slow rate of growth. |
| **Texas Redbud** (*Cercis canadensis* v. *texensis*) | Semi-Arid<br>Desert<br>Moist Subtropical | 20 × 20 ft (6.1 × 6.1 m) | -10°F (-23°C) | Small, deciduous tree with lush-green foliage. Pink/purple blooms appear in spring followed by seedpods. Leaves turn yellow in fall. |
| **Desert Willow** (*Chilopsis linearis*) | Semi-Arid<br>Desert<br>Moist Subtropical | 20–30 × 20–30 ft (6.1–9.1 × 6.1–9.1 m) | -10°F (-23°C) | Deciduous tree with bright-green, narrow leaves. Pink to burgundy flowers are produced in summer. Seedless cultivars available. |
| **Chitalpa** (× *Chitalpa tashkentensis*) | Coastal<br>Semi-Arid<br>Desert<br>Moist Subtropical | 15 × 15 ft (4.6 × 4.6 m) | -10°F (-23°C) | Small, deciduous tree with dark-green foliage and lovely white to light-pink flowers that bloom in summer. There are seedless cultivars available. |
| **Texas Olive, Mexican Olive** (*Cordia boissieri*) | Semi-Arid<br>Desert<br>Moist Subtropical<br>Tropical | 25 × 25 ft (7.6 × 7.6 m) | 20°F (-7°C) | Small tree with dark-green, velvety leaves. White flowers appear in summer followed by a small, green fruit. The foliage is evergreen. |
| **Turkish Hazel** (*Corylus colurna*) | Coastal<br>Moist Subtropical<br>Moist Continental<br>Highlands | 40–50 × 15–20 ft (12.2–15.2 × 4.6–6.1 m) | -30°F (-34°C) | Upright, deciduous tree with a teardrop shape. Its lush green foliage and strong wood makes it resistant to damage from wind or snow. |
| **Texas Mountain Laurel** (*Dermatophyllum secundiflorum* syn. *Sophora secundiflora*) | Semi-Arid<br>Desert<br>Moist Subtropical | 10–20 × 15 ft (3–6.1 × 4.6 m) | 10°F (-12°C) | Small, slow-growing tree with lush, evergreen foliage. Fragrant, purple flowers that smell like grape soda appear in spring. It is also grown as a tall shrub. |

## Trees, cont.

| NAME | REGION | MATURE SIZE | COLD HARDINESS | NOTES |
|---|---|---|---|---|
| **Kentucky Coffee Tree** (*Gymnocladus dioicus*) | Moist Subtropical<br>Moist Continental | 60–100 × 40–55 ft (18.3–30.4 × 12.2–16.8 m) | -40°F (-40°C) | Deciduous tree with lush-green foliage and rough gray bark. Available in male (no seed-pods) or female trees (has fragrant flowers and seedpods). |
| **Eastern Red Cedar** (*Juniperus virginiana*) | Moist Subtropical<br>Moist Continental | 30–90 ft (9.1–27.4 m) | -45°F (-43°C) | Aromatic, scalelike leaves grace the limbs of this slow-growing evergreen tree. Foliage color can vary from gray to dark green with a number of cultivars available. |
| **Lacebark Pine** (*Pinus bungeana*) | Semi-Arid<br>Moist Subtropical<br>Moist Continental<br>Highlands | 30–50 × 20–35 ft (9.1–15.2 × 6.1–10.6 m) | -25°F (-32°C) | Slow-growing evergreen tree with medium-green needlelike foliage. The trunk is colorful with shades of gray, green, and rust. |
| **'Red Push' Pistache** (*Pistacia* × 'Red Push') | Semi-Arid<br>Desert<br>Moist Subtropical | 25–40 × 20–30 ft (7.6–12.2 × 6.1–9.1 m) | 0°F (-18°C) | Deciduous tree with lush green leaves. Prized for its traditional tree shape and colorful autumn foliage. |
| **Honey Mesquite** (*Prosopis glandulosa*) | Semi-Arid<br>Desert<br>Moist Subtropical | 25 × 30–40 ft (7.6 × 9.1–12.2 m) | 0°F (-18°C) | Deciduous, fast-growing tree with curving branches and bright-green foliage. Best when grown as a multi-trunk tree. |
| **Hop Tree** (*Ptelea trifoliata*) | Moist Subtropical<br>Moist Continental | 10–20 × 15–20 ft (3–6.1 × 4.6–6.1 m) | -35°F (-37°C) | Deciduous, small, multi-trunk tree with lush green foliage that turns yellow in autumn. Attracts birds and butterflies. |
| **Escarpment Live Oak** (*Quercus fusiformis*) | Desert<br>Moist Subtropical | 50 × 50 ft (15.2 × 15.2 m) | -10°F (-23°C) | Deep green foliage that is evergreen in mild-winter climates and deciduous in cold winter regions. Has a slow rate of growth. |
| **Bur Oak** (*Quercus macrocarpa*) | Moist Subtropical<br>Moist Continental | 60–80 × 60–80 ft (18.3–24.4 × 18.3–24.4 m) | -40°F (-40°C) | Large oak with dark-green foliage. Mature specimens produce large acorns. In moister soils, it can reach heights over 100 feet (30 m). |
| **Chinese Elm** (*Ulmus parvifolia*) | Coastal<br>Semi-Arid<br>Desert<br>Moist Subtropical<br>Moist Continental<br>Highlands | 35–50 × 35–50 ft (10.6–15.2 × 10.6–15.2 m) | -20°F (-29°C) | Shade tree with wide canopy that is ever-green in mild winter climates. The trunk has mottled coloring with gray, red-brown, and tan colors. |
| **Chaste Tree, Vitex** (*Vitex agnus-castus*) | Semi-Arid<br>Desert<br>Moist Subtropical<br>Tropical | 25 × 25 ft (7.6 × 7.6 m) | -10°F (-23°C) | Slow-growing deciduous shrub trained as a tree. Spikes of purple flowers appear in summer. |

**After trees, shrubs are the second most essential plants in the landscape.** They come in different sizes and shapes and perform many other functions. Shrubs add beauty to the garden, serving functions such as a privacy screen, blocking objectionable views, wind reduction, shade, and wildlife habitat. Additionally, tall shrubs are a good shade substitute for smaller spaces where trees won't fit.

## Shrubs

| NAME | REGION | MATURE SIZE | COLD HARDINESS | NOTES |
|---|---|---|---|---|
| **Bougainvillea** (*Bougainvillea* spp.) | Semi-Arid<br>Desert<br>Moist Subtropical<br>Tropical | 5–10 × 5–8 ft (1.5–3 × 1.5–2.4 m) | 25°F (-4°C) (root-hardy to 10°F [-12°C]) | Tropical shrub with lush green leaves that hide thorns. Colorful bracts range from red, magenta, orange to white depending on species. Frost damage occurs at 32°F (0°C). |
| **Woolly Butterfly Bush** (*Buddleja marrubiifolia*) | Semi-Arid<br>Desert | 5 × 5 ft (1.5 × 1.5 m) | 10°F (-12°C) | Soft gray foliage, orange flowers appear off and on throughout the year. Attracts butterflies. |
| **Pride of Barbados, Red Bird-of-Paradise** (*Caesalpinia pulcherrima*) | Semi-Arid<br>Desert<br>Moist Subtropical<br>Tropical | 6–10 × 6–10 ft (1.8–3 × 1.8–3 m) | 10°F (-12°C) | Semi-deciduous shrub with lush-green lacy foliage. Colorful orange/red/orange flowers appear in summer. |
| **Pink Fairy Duster** (*Calliandra eriophylla*) | Semi-Arid<br>Desert<br>Moist Subtropical | 3 × 3 ft (91 × 91 cm) | 10°F (-12°C) | Small, rounded shrub with small sage-green leaves. Pink puff ball flowers appear in spring and in early fall. Attracts hummingbirds. |
| **'Blue Mist' Spirea** (*Caryopteris* x *clandonensis* 'Blue Mist') | Moist Subtropical<br>Moist Continental<br>Highlands | 2–3 × 2–3 ft (61–91 × 61–91 cm) | -15°F (-26°F) | Deciduous shrub with aromatic, medium-green leaves. Blue flowers appear in summer and attract bees and butterflies. |
| **Victoria California Lilac** (*Ceanothus thyrsiflorus* 'Victoria') | Coastal<br>Semi-Arid<br>Moist Subtropical | 9 × 10–12 ft (2.7 × 3–3.7 m) | 5°F (-15°C) | Large shrub with dark green foliage. Fragrant blue flowers appear in spring, attracting butterflies and hummingbirds. |
| **Smoke Bush** (*Cotinus coggygria*) | Coastal<br>Semi-Arid<br>Moist Subtropical<br>Moist Continental | 10–15 × 10–15 ft (3–4.6 × 3–4.6 m) | -30°F (-34°C) | Large, deciduous shrub with variable leaf color from deep green to reddish purple. Produces airy, smokey-pink blooms that deepen in color. Fall foliage ranges from yellow, orange to red. |
| **Hop Bush, Hopseed Bush** (*Dodonaea viscosa*) | Semi-Arid<br>Desert<br>Moist Subtropical | 5–12 × 6–10 ft (1.5–3.7 × 1.8–3 m) | 15°F (-9°C) | Large shrub with medium-green foliage. Makes an excellent hedge or screen. |
| **Turpentine Bush** (*Ericameria laricifolia*) | Semi-Arid<br>Desert | 3 × 3 ft (91 × 91 cm) | -10°F (-23°C) | Small shrub with lush green, aromatic foliage with needle-shaped leaves. Golden yellow flowers appear in late summer into fall. |

## Shrubs, cont.

| NAME | REGION | MATURE SIZE | COLD HARDINESS | NOTES |
|---|---|---|---|---|
| **Chuparosa** (*Justicia californica*) | Semi-Arid<br>Desert<br>Moist Subtropical | 3–5 × 5 ft (91 cm–1.5 m × 1.5 m) | 20°F (-7°C) | Sprawling shrub with bright-green foliage and gray stems. Red tubular flowers appear year-round and attract hummingbirds. |
| **Texas Sage** (*Leucophyllum* spp.) | Semi-Arid, Desert<br>Moist Subtropical<br>Tropical | 5–6 × 5–8 ft (1.5–1.8 × 1.5–2.4 m) | 10°F (-12°C) | Large shrub with sage-green to gray-blue foliage. Increased humidity spring through summer stimulates flowering in white or varying shades of purple. |
| **Heavenly Bamboo** (*Nandina domestica*) | Coastal, Semi-Arid<br>Desert<br>Moist Subtropical<br>Tropical | 4–8 × 2–4 ft (1.2–2.4 m × 61 cm–1.2 m) | -10°F (-23°C) | Evergreen shrub prized for its bright-green foliage that turns maroon at the tips in fall. White flowers in spring are followed by red berries that birds enjoy. |
| **Littleleaf Mock Orange** (*Philadelphus microphyllus*) | Semi-Arid<br>Moist Subtropical<br>Highlands | 4–6 × 4–6 ft (1.2–1.8 × 1.2–1.8 m) | -20°F (-29°C) | Drought-tolerant species of popular, deciduous shrub with attractive, peeling bark. Lush-green foliage is covered with fragrant, white flowers in spring. |
| **Chaparral Sage** (*Salvia clevelandii*) | Semi-Arid<br>Desert | 5 × 5 ft (1.5 × 1.5 m) | 10°F (-12°C) | Shrub with aromatic sage-green leaves. Blue-purple blooms appear in spring. Attracts pollinators. Plant in full sun (afternoon shade in desert gardens). |
| **Rosemary** (*Salvia rosmarinus* syn. *Rosmarinus officinalis*) | Semi-Arid, Desert<br>Moist Subtropical<br>Moist Continental (Annual)<br>Highlands (Annual)<br>Tropical | 4–6 × 4–5 ft / 2 × 4 ft (1.2–1.8 × 1.2–1.5 m / 61 cm × 1.2 m) | 10°F (-12°C) | Needlelike, aromatic leaves that are dark green. Available in shrub or groundcover forms. The foliage is edible and used as an herb. |
| **Russian Sage** (*Salvia yangii* syn. *Perovskia atriplicipolia*) | Coastal<br>Semi-Arid, Desert<br>Moist Continental<br>Moist Subtropical | 3–4 × 3–4 ft (91 cm–1.2 m × 91 cm–1.2 m) | -30°F (-34°C) | Shrubby perennial with small, silvery-gray leaves. Has an open spiky shape, and light purple flowers appear in summer. |
| **Jojoba** (*Simmondsia chinensis*) | Semi-Arid<br>Desert | 5 × 10 ft (1.5 × 3 m) | 10°F (-12°C) | Oval-shaped, gray-green leaves. Cosmetics are made from the wax of the female plant's small brown fruit. |
| **Yellow Bells** (*Tecoma stans*) | Semi-Arid<br>Desert<br>Moist Subtropical<br>Tropical | 6–8 × 8 ft (1.8–2.4 × 2.4 m) | 20°F (-7°C) | Large shrub with lush-green foliage and yellow, trumpet-shaped flowers. Available in other varieties with orange to red-colored flowers. *T. stans* v. *angustata* is a more cold-tolerant variety. |
| **Shrubby Germander** (*Teucrium fruticans*) | Semi-Arid<br>Desert | 3–4 × 5 ft (91 cm–1.2 m × 1.5 m) | 10°F (-12°C) | Leaves are silvery-blue on top and white on the bottom. Blue flowers appear in spring through fall (winter-spring in low desert regions). |

**Smaller plants such as groundcovers and perennials** provide vital interest at lower levels. Additionally, they can create a "living mulch," providing shade and cooling to roots during the summer months. These shorter plants don't generally live as long as trees and shrubs. Yet, they are inexpensive, easy to replace, and an excellent substitute for thirsty lawns—plant groundcovers and perennials in groups of three or more for maximum impact.

## Groundcovers and Perennials

| NAME | REGION | MATURE SIZE | COLD HARDINESS | NOTES |
|---|---|---|---|---|
| **Yarrow** (*Achillea millefolium*) | Coastal, Semi-Arid<br>Moist Subtropical<br>Moist Continental<br>Highlands<br>Tropical | 2–3 × 2–3 ft (61–91 × 61–91 cm) | -40°F (-40°C) | Perennial with fernlike green leaves. White tiny flowers appear in clusters throughout summer. Cultivars include pink, yellow, and other flower colors. |
| **Hummingbird Mint** (*Agastache* spp.) | Coastal<br>Semi-Arid, Desert<br>Moist Subtropical<br>Moist Continental<br>Highlands | 1–6 × 2–3 ft (30 cm–1.8 m × 61–91 cm) | -40°F (-40°C) | Upright perennial with medium-green, aromatic foliage. Flowering spikes appear in summer. *Cold hardiness, size, and bloom color depends on the variety. |
| **Chocolate Flower** (*Berlandiera lyrata*) | Semi-Arid, Desert<br>Moist Subtropical | 1 × 2 ft (30 × 61 cm) | -30°F (-34°C) | Yellow-flowering perennial with gray-green foliage. The flowers smell like chocolate. |
| **Jupiter's Beard** (*Centranthus ruber*) | Coastal, Semi-Arid<br>Moist Subtropical<br>Moist Continental | 1.5–2 × 2–3 ft (46–61 × 61–91 cm) | -15°F (-26°C) | Clumping perennial with fleshy, medium-green leaves. Spikes of red/pink flowers bloom in spring that attract birds and butterflies. |
| **Snow in Summer** (*Cerastium tomentosum*) | Semi-Arid<br>Moist Continental<br>Highlands | 6–12 × 8–12 in (15–30 × 20–30 cm) | -40°F (-40°C) | Spreading groundcover with gray-green foliage. White daisylike flowers appear in summer. |
| **Large-Flowered Tickseed** (*Coreopsis grandiflora*) | Coastal, Semi-Arid<br>Moist Subtropical<br>Moist Continental<br>Highlands | 1.5–2.5 × 1–1.5 ft (46–76 × 30–46 cm) | -30°F (-34°C) | Golden yellow flowers cover this popular perennial in summer. The foliage consists of narrow, lush-green leaves. Rabbit/deer resistant. |
| **Purple Coneflower** (*Echinacea purpurea*) | Coastal, Semi-Arid<br>Moist Subtropical<br>Moist Continental<br>Highlands | 2–4 × 1.5–2 ft (61 cm–1.2 m × 46–61 cm) | -40°F (-40°C) | Purple daisylike flowers with prominent orange centers are borne on stiff stems over dark green foliage. Attracts butterflies and hummingbirds. |
| **Blanket Flower** (*Gaillardia × grandiflora*) | Coastal, Semi-Arid<br>Desert<br>Moist Subtropical<br>Moist Continental<br>Tropical | 1–2 × 1–2 ft (31–61 × 31–61 cm) | -40°F (-40°C) | Colorful perennial with medium-green foliage. Flower color is shades of red, orange, and yellow and flowers appear in summer. |
| **Prairie Feather** (*Liatris spicata*) | Moist Subtropical<br>Moist Continental | 1.5 × 1.5 ft (46 × 46 cm) | -40°F (-40°C) | Small groundcover with needle-shaped foliage that is dark green. Lavender flowering spikes appear in late summer. |

## Groundcovers and Perennials, cont.

| NAME | REGION | MATURE SIZE | COLD HARDINESS | NOTES |
|---|---|---|---|---|
| **Blackfoot Daisy** (*Melampodium leucanthum*) | Semi-Arid, Desert<br>Moist Subtropical<br>Moist Continental | 1 × 2 ft (30 × 61 cm) | -20°F (-29°C) | Softly mounding perennial with dark-green leaves. White flowers with yellow centers appear throughout the growing season. |
| **Colorado Four O'Clock** (*Mirabilis multiflora*) | Semi-Arid<br>Moist Subtropical<br>Moist Continental<br>Highlands | 2 × 3 ft (61 × 91 cm) | -20°F (-29°C) | Low-growing groundcover with blue-green foliage. Magenta flowers appear in summer that last a day. Attracts butterflies and hummingbirds. |
| **Penstemon** (*Penstemon* spp.) | Semi-Arid, Desert<br>Moist Subtropical<br>Moist Continental<br>Highlands | 1 × 2 ft (30 × 61 cm) | -30°F to 25°F (-34°C to -4°C) (depending on species) | Flowering spikes appear over clumping foliage, which attract hummingbirds. Blooming occurs in spring (summer in cooler regions). Cold hardiness and bloom color dependent on the species. |
| **Prairie Coneflower, Mexican Hat Plant** (*Ratibida columnifera*) | Semi-Arid, Desert<br>Moist Subtropical<br>Moist Continental | 1–3 × 1.5 ft (30–91 × 46 cm) | -30°F (-34°C) | Perennial with a wildflower appearance. Yellow to rust-red flowers with raised centers appear in spring/summer. Foliage finely textured and medium green in color. |
| **Black-Eyed Susan** (*Rudbeckia hirta*) | Coastal, Semi-Arid<br>Moist Subtropical<br>Moist Continental<br>Highlands, Tropical | 2–3 × 2 ft (61–91 × 61 cm) | -30°F (-34°C) | Deep yellow daisylike flowers with prominent brown centers appear in summer to early fall over dark-green foliage. Attracts birds and butterflies. |
| **Autumn Sage** (*Salvia greggii*) | Coastal, Semi-Arid<br>Desert<br>Moist Subtropical<br>Moist Continental<br>Tropical | 2–3 × 2–3 ft (61–91 × 61–91 cm) | 0°F (-18°C) | Shrubby perennial with medium-green, aromatic foliage. Red flowers attract hummingbirds. Pink, purple, and white cultivars available. |
| **Mexican Bush Sage** (*Salvia leucantha*) | Coastal<br>Semi-Arid, Desert<br>Moist Subtropical<br>Tropical | 3–4 × 3 ft (91 cm–1.2 m × 91 cm) | 18°F (-8°C) | Perennial with a shrubby growth habit. Spikes of purple/white flowers appear in spring and summer, attracting hummingbirds. |
| **Lemon Coral Stonecrop** (*Sedum mexicanum* 'Lemon Coral') | Coastal<br>Semi-Arid<br>Moist Subtropical<br>Moist Continental | 4 × 12–24 in (10 × 30–61 cm) | -20°F (-29°C) | Extremely low-growing, succulent groundcover with medium-green foliage. Can handle minimal foot traffic. |
| **Lamb's Ear** (*Stachys byzantina*) | Semi-Arid<br>Moist Subtropical<br>Moist Continental<br>Highlands | 6–12 × 12–16 in (15–30 × 30–41 cm) | -30°F (-34°C) | Groundcover prized for its soft gray-green foliage. The leaves have a velvety texture. They rarely flower and the blooms aren't very notable. |
| **Angelita Daisy, Four Nerve Daisy** (*Tetraneuris acaulis*) | Semi-Arid<br>Desert<br>Moist Subtropical<br>Highlands | 4–6 × 6–8 in (10–15 × 15–20 cm) | -30°F (-34°C) | Petite perennial with dark-green, narrow leaves. Golden-yellow flowers appear in spring, summer, and frost-free winters. |

**Vines transform the vertical spaces around our homes,** covering them in shades of green and often treating us to a colorful display of blooms. They are a wonderful option for narrow areas where you want coverage, yet a shrub won't fit. Many vines need support to grow upward, which a fence or trellis can provide. It's important to keep vines from outgrowing their allotted space, or you might find your favorite tree or shrub overcome with growth.

## Vines

| NAME | REGION | MATURE SIZE | COLD HARDINESS | NOTES |
|---|---|---|---|---|
| **Crossvine** (*Bignonia capreolata*) | Semi-Arid, Desert<br>Moist Subtropical | 30 × 30 ft (9.1 × 9.1 m) | -10°F (-23°C) | Vine with lush-green foliage that climbs up a trellis or other support. Orange/pink flowers appear in spring to early summer. |
| **Arizona Grape Ivy** (*Cissus trifoliata*) | Semi-Arid<br>Desert<br>Moist Subtropical | 15–30 × 5–10 ft (4.6–9.1 × 1.5–3 m) | 20°F (-7°C) | The foliage of this vine creates a medium-green backdrop as it climbs upward on a trellis. Deciduous in cold climates. Insignificant flowers develop into black fruit. Attracts bees, butterflies, and birds. |
| **Common Hop** (*Humulus lupulus*) | Coastal, Semi-Arid<br>Moist Subtropical<br>Moist Continental<br>Highlands | 15–20 × 3–6 ft (4.6–6.1 m × 91 cm–1.8 m) | -20°F (-29°C) | Vine with attractive foliage that produces green flowers that develop into cones that are used for flavoring beer. It dies back to the ground in winter but grows back from the roots in spring. |
| **Honeysuckle** (*Lonicera* spp.) | Coastal, Semi-Arid<br>Desert<br>Moist Subtropical<br>Moist Continental<br>Tropical | 5–20 × 5–15 ft (1.5–6.1 × 1.5–4.6 m) | -30°F (-34°C) | Clusters of tubular flowers provide lovely fragrance through summer. Bloom color depends on the variety. Attracts pollinators. |
| **Purple Passionflower** (*Passiflora incarnata*) | Coastal, Semi-Arid<br>Desert<br>Moist Subtropical<br>Moist Continental<br>Tropical | 10–20 × 5–10 ft (3–6.1 × 1.5–3 m) | -5°F (-21°C) | Exotic, purple flowers develop into oval, globular fruits that are edible. The green leaves add warm-season interest. Attracts butterflies. |
| **Mexican Flame Vine** (*Pseudogynoxys chenopodioides* syn. *Senecio confusus*) | Semi-Arid<br>Desert<br>Moist Subtropical<br>Tropical | 6–12 × 4–6 ft (1.8–3.7 × 1.2–1.8 m) | 20°F (-7°C) | Evergreen vine with lush green foliage with a long bloom season during the warm season. Flowers are a striking orange color. Attracts bees and butterflies. |
| **Lady Banks' Rose** (*Rosa banksiae*) | Coastal<br>Semi-Arid<br>Desert<br>Moist Subtropical | 10–20 × 12–15 ft (3–6.1 × 3.7–4.6 m) | 10°F (-12°C) | Thornless rose vine that blooms once a year in spring. Flowers are white or yellow depending on the variety. Needs a trellis or other support. |
| **Star Jasmine** (*Trachelospermum jasminoides*) | Semi-Arid<br>Desert<br>Moist Subtropical<br>Tropical | 8–20 × 6–15 ft (2.4–6.1 × 1.8–4.6 m) | 10°F (-12°C) | Deep green leaves climb upward on dark brown stems. Fragrant, star-shaped flowers appear in spring. |

**Grasses add lovely texture to the garden** and are very versatile in the look of your landscape. I like to plant them in groups of three or more near boulders. Alternatively, they are useful in more formal, contemporary designs when planted in rows or against a building or used to create a grasslike hedge.

## Ornamental Grasses

| NAME | REGION | MATURE SIZE | COLD HARDINESS | NOTES |
|---|---|---|---|---|
| **Blue Grama** (*Bouteloua gracilis*) | Coastal<br>Semi-Arid<br>Desert<br>Moist Subtropical<br>Moist Continental<br>Highlands | 2 × 2 ft (61 × 61 cm) | -30°F (-34°C) | Fine-textured grass, medium-green leaves form an upright, arching growth habit. Blonde "flowers" that look like eyebrows form at the tips throughout the growing season. |
| **Giant Feather Grass** (*Celtica gigantea* syn. *Stipa gigantea*) | Coastal<br>Semi-Arid<br>Moist Subtropical<br>Moist Continental<br>Highlands | 5–8 × 2 ft (1.5–2.4 m × 61 cm) | -20°F (-29°C) | Large grass with tan seed heads that appear on arching stems over finely-textured foliage, which adds summer and fall interest. |
| **Blue Fescue** (*Festuca glauca*) | Coastal<br>Semi-Arid<br>Desert<br>Moist Subtropical<br>Moist Continental<br>Highlands | 1 × 1 ft (30 × 30 cm) | -30°F (-34°C) | Semi-evergreen ornamental grass with finely textured, blue-gray leaves. Its short height makes it an excellent groundcover option. Flowers in early summer. |
| **Blue Oats Grass** (*Helictotrichon sempervirens*) | Coastal<br>Semi-Arid<br>Desert<br>Moist Subtropical<br>Moist Continental<br>Highlands | 2 × 2 ft (61 × 61 cm) | -30°F (-34°C) | Evergreen grass with blue-gray, narrow leaves. Flower spikes that resemble oats appear in summer and turn tan in autumn. |
| **Pink Muhly Grass** (*Muhlenbergia capillaris*) | Semi-Arid<br>Desert<br>Moist Subtropical<br>Moist Continental<br>Highlands<br>Tropical | 3 × 3 ft (91 × 91 cm) | -10°F (-23°C) | Finely textured grass with medium-green leaves. Burgundy plumes appear in late summer and fall that turn tan in winter. |
| **Deer Grass** (*Muhlenbergia rigens*) | Coastal<br>Semi-Arid<br>Desert | 4–5 × 4–5 ft (1.2–1.5 × 1.2–1.5 m) | -10°F (-23°C) | Tan flower spikes appear among finely textured gray-green leaves that appear in late summer into fall. |

**Succulents, including cacti, can be grown in a range of different climates.** However, well-drained soil is crucial to their survival. For areas where rainfall is plentiful and moist soil is typical, locate succulents in higher areas of the garden or plant them in containers where soil can completely dry out for several days between watering. Succulents are adapted to dry soil conditions and are often killed by overwatering. The frequency of watering is dependent on your local climate and the season. Your best resource for succulent care is to ask your local nursery professional for specific care for your region.

## Succulents

| NAME | REGION | MATURE SIZE | COLD HARDINESS | NOTES |
|---|---|---|---|---|
| **New Mexico Agave** (*Agave parryi* v. *neomexicana*) | Coastal<br>Semi-Arid<br>Desert<br>Highlands | 1.5 × 2 ft (46 × 61 cm) | -20°F (-29°C) | Blue-gray leaves with beautiful leaf imprints create an attractive rosette. Each leaf is edged with small maroon "teeth" and a single spine at the tip. |
| **Weber Agave** (*Agave weberi*) | Semi-Arid<br>Desert<br>Tropical | 5 × 6–8 ft (1.5 × 1.8–2.4 m) | 10°F (-12°C) | Long, gray-green leaves create an attractive spiky accent. Best used in areas with plenty of room for mature growth. |
| **'Blue Elf' Aloe** (*Aloe* × 'Blue Elf') | Semi-Arid<br>Desert<br>Moist Subtropical | 18 in × 2 ft (46 × 61 cm) | 20°F (-7°C) | Small aloe with gray-green upright leaves. Orange, tubular flowers appear in spring and attract hummingbirds. |
| **Desert Spoon** (*Dasylirion wheeleri*) | Coastal<br>Semi-Arid<br>Desert<br>Moist Subtropical | 6 × 6 ft (1.8 × 1.8 m) | 10°F (-12°C) | Large succulent with narrow spiky, bright-gray narrow leaves that fan outward. Leaves have serrated edges with tiny "teeth." |
| **Ice Plant** (*Delosperma cooperi*) | Coastal<br>Semi-Arid<br>Desert<br>Moist Subtropical<br>Moist Continental | 3–6 × 12–24 in (8–15 × 30–61 cm) | -10°F (-23°C) | Succulent groundcover with narrow, medium-green leaves. Daisylike, deep-pink flowers appear in summer. |
| **Candelilla** (*Euphorbia antisyphilitica*) | Desert | 1–2 × 2–3 ft (30–61 × 61–91 cm) | 10°F (-12°C) | Upright succulent with gray-green stems with a clumping growth habit. Tiny, pale-pink flowers may appear for a few days in spring. Thrives in full sun. |
| **Gopher Plant** (*Euphorbia rigida* syn. *E. biglandulosa*) | Coastal<br>Semi-Arid<br>Desert<br>Moist Subtropical<br>Moist Continental<br>Highlands | 2–3 × 3 ft (61–91 × 91 cm) | -20°F (-29°C) | Succulent groundcover with narrow, gray-green leaves arranged along fleshy stems. Chartreuse flowers appear in spring. |

## Succulents, cont.

| NAME | REGION | MATURE SIZE | COLD HARDINESS | NOTES |
|---|---|---|---|---|
| **Red Yucca** (*Hesperaloe parviflora*) | Coastal<br>Semi-Arid<br>Desert<br>Moist Subtropical<br>Moist Continental<br>Highlands | 3 × 5 ft (91 cm × 1.5 m) | -20°F (-29°C) | Dark-green leaves that resemble an ornamental grass. Coral flowers appear in spring and summer, attracting hummingbirds. |
| **Golden Barrel** (*Kroenleinia* syn. *Echinocactus grusonii*) | Coastal<br>Semi-Arid<br>Desert<br>Moist Subtropical | 2–3 × 2.5 ft (61–91 × 76 cm) | 20°F (-7°C) | Globe-shaped cactus with golden-yellow, curved spines that cover its medium-green surface. |
| **Beavertail Prickly Pear** (*Opuntia basilaris*) | Coastal<br>Semi-Arid<br>Desert | 2 × 3 ft (61 × 91 cm) | 0°F (-18°C) | Compact cactus with rounded, blue-gray pads. Vibrant pink flowers are produced in late spring/early summer. |
| **Hardy Spineless Prickly Pear** (*Opuntia cacanapa* 'Ellisiana') | Coastal<br>Semi-Arid<br>Desert<br>Moist Subtropical | 3–4 × 6 ft (91 cm–1.2 m × 1.8 m) | 0°F (-18°C) | Cactus with medium-green, paddle-shaped segments. Unlike most other prickly pear cacti, this has a fairly smooth surface without sharp spines. |
| **Lemon Coral Stonecrop** (*Sedum mexicanum* 'Lemon Coral') | Coastal<br>Semi-Arid<br>Moist Subtropical<br>Moist Continental<br>Highlands | 3–10 × 10–14 in (8–25 × 25–36 cm) | 0°F (-18°C) | Succulent groundcover with yellow-green leaves. Small star-shaped, yellow blooms appear in summer. Makes a wonderful container plant. |
| **'Autumn Joy' Sedum** (*Sedum spectabile* 'Autumn Joy') | Semi-Arid<br>Moist Subtropical<br>Moist Continental<br>Highlands | 1.5–2 × 1.5–2 ft (46–61 × 46–61 cm) | -40°F (-40°C) | Cold-hardy succulent that resembles a perennial. Fleshy, light-green leaves are topped with pinkish-red blooms in late summer to fall. Attracts butterflies. |
| **Hens and Chicks** (*Sempervivum tectorum*) | Coastal<br>Semi-Arid<br>Moist Subtropical<br>Moist Continental<br>Highlands | 6–12 × 6–18 in (15–30 × 15–46 cm) | -30°F (-34°C) | Light-green succulent with leaves arranged in a rosette pattern. The tips of leaves may turn purple. Reddish/purple flowers appear in summer on 1-foot (30 cm) tall stems. |
| **Plains Yucca** (*Yucca glauca*) | Semi-Arid<br>Desert<br>Moist Subtropical<br>Moist Continental<br>Highlands | 3–4 × 3–4 ft (91 cm–1.2 m × 91 cm–1.2 m) | -30°F (-34°C) | Spiky succulent with narrow, gray-green leaves. In summer, greenish-white flowers bloom on a 4-foot (1.2 m) stem that attract butterflies. |

## Where to Find Plants for a Water-Smart Garden

When it comes to buying plants, there are many different options available. However, some sources are better than others. It's vital to purchase plants that will succeed in your climate. Temperature, humidity, and rainfall are factors regarding whether a plant will flourish in your garden. I recommend the following places for getting plants, listed in order of preference:

### *Botanical Garden Plant Sale*

Botanical garden plant sales are your best and most trusted place to get plants. The gardens aren't just an invaluable resource for seeing what plants do well in your climate; they do an excellent job of selecting plants to sell that will do well where you live, many of which are on display at the garden. They also often offer plants that are hard to find in the retail nursery setting. Plant experts are usually present to answer your specific questions. Some gardens hold seasonal plant sales only, while others have plants for sale year-round. Proceeds often go to help fund the botanical garden.

### *Local Nursery/Garden Center*

Many people have a favorite nursery where they find plants, garden supplies, and something else essential—reliable advice tailored for your region. I recommend local nurseries and not those that are national. A local nursery professional is one of your best resources when choosing the right plants and how to maintain them properly. A nursery professional usually has completed a certification program, and it's okay to ask them if they have. Your local nursery may or may not provide a guarantee for plants, so it's important to ask.

### *Big Box Stores*

Big box stores are hard to beat when it comes to convenience and lower prices. However, while you can find suitable plants in these stores, you can run into potential problems. These stores can carry some plants that don't do well in your climate. Sadly, this is quite common, and it can be challenging for the consumer (you) to tell which ones will do well and what won't. Therefore, the advice provided by employees may not always be accurate. Saving money means nothing if the plant dies or struggles to survive. For some people, a big box store may be their only option for buying plants. You can purchase plants at these stores, but do your homework first and make sure you select plants that will grow in your region. Lastly, avoid impulse buys. Big box stores often place their showiest plants toward the entrance to tempt you—make sure it is suitable for the climatic conditions where you live.

Plants available at botanical garden plant sales are reliable sources for plants with region-specific guidelines.

## Research Plant Choices Before Buying

Before heading out to buy plants for your garden, take time to research them ahead of time. It's important to know their mature size, water requirements, preferred sun exposure, hot and cold temperature tolerances, the soil they prefer (well-drained, moist, etc.), and the type of maintenance they will require. Thankfully, there are a lot of resources available to you to find out these characteristics so you can select the right plants. Many botanical gardens have sections with low-water plants you can explore. Visit local nurseries (without buying at first) to see what's available and ask questions. Online searches can provide a wealth of information. When searching for plants online, ensure you include your region in the search parameters to get information specific to your growing conditions. Regional gardening books with a plant profile section can be invaluable in finding the right plants. Finally, observe the plants in your neighborhood, town, city, etc. Note what plants you like that are flourishing, take pictures of them, and visit your local nursery to help you identify them.

Plant selection is crucial to creating an outdoor space that uses water efficiently. Avoid using plants that require a lot of water or restrict thirsty plants to a smaller area, such as a container garden. Most of all, have fun exploring the many drought-resistant plants that will add beauty around your home!

Spanish lavender (*Lavandula stoechas*) thrives in this coastal Oregon garden in well-drained soil.

## The Best Time to Plant for Water Efficiency

When we add our plants affects how much water they will need to become established. The ideal planting season is in the fall for water efficiency, providing plants with up to nine months to grow their roots before the stress of summer, when water needs are at their highest. The more roots a plant has, the better its ability to take up water. Newly planted plants have small root systems and require frequent applications of water to combat the soil drying out quickly around their roots. If you plant in summer, the weather is warmer, and plants lose more water to evaporation, which puts more stress on plants, requiring more frequent watering. Focus on adding plants in the fall and at least six weeks before the first frost for cold winter regions. If you want to add plants that are susceptible to damage from winter cold, wait until spring to install them or provide protection from winter cold.

This young plant has a root system that has been limited by the nursery container, until now. Once planted, it will grow a more extensive root system over time with some extra watering to help it become established.

CHAPTER 3

# Where and How to Plant for Water Efficiency

**When it comes to creating a garden** or making changes to the one you already have, we base our decisions about plant placement on the function or beauty of the plants. However, we also need to include another factor in our decision-making process: how much water plants will require. Climatic conditions that plants are exposed to influence a plant's water needs around your garden.

## Where You Place Your Plants Affects Their Water Use

The amount of sun a plant receives affects how much water it needs. The sun draws water out of a plant's leaves, which means the more sun, the more potential water loss for the plant. The sun's effects on a plant's water usage means that plants grown in full sun will need more water than those grown in shade unless they have particular adaptations. Exposure to wind also will affect how much water a plant needs. Sun and wind should influence how we approach designing our outdoor space. Plants also can function to modify the climate around your home, creating areas of shade, reducing wind, and making the most of rainfall or supplemental irrigation.

A waterwise landscape in Utah with purple coneflower (*Echinacea purpurea*), blue fescue (*Festuca glauca*), and black-eyed Susan (*Rudbeckia* spp.).

Planting drought-tolerant Texas sage (*Leucophyllum* spp.) in an area with full sun and reflected heat.

## WHEN A PLANT LABEL ISN'T ALWAYS ACCURATE

When shopping for plants, it's important to note that "full sun" on the plant label doesn't mean that the plant will thrive in all climates in full sun. In desert and tropical regions, the sun's intensity is much greater and may be too much for plants. Observe plants around your neighborhood or botanical garden to see which do best in the sun, and consult with your local nursery professional.

### Why Proper Sun Exposure Is Such a Big Deal

You are likely familiar with the concept of sun and shade in the garden and how important it is to match a plant's preference for a particular amount of light. Some plants do best in full sun, while others prefer the lower light shade conditions. Sun exposure is more than just "sun" and "shade." Several exposures lie in between, which are essential to know when it comes to matching plants with their preferred amount of light:

- **Reflected Heat**—sunny areas that receive reflected sun and heat from nearby walls, sidewalks, or another type of pavement.
- **Full Sun**—receive sun (no shade) for at least six hours or more daily.
- **Part Sun/Filtered Sun**—areas that experience a mixture of sun and shade.
- **Shade**—locations that receive no direct sun; some may have higher light levels while others are rather dark.

Some plants have adaptations (natural plant sun protection) to allow them to handle sunny conditions without losing too much water. These include tiny hairs covering the foliage, a thick coating on the leaf, and smaller leaves. Plants not adapted to full sun lack these characteristics and are susceptible to losing a lot of water through their leaves. Additionally, they also can suffer from sunburned foliage too. Sadly, many people experience frustration and disappointment when they don't match plants with the correct exposure. Be sure to read the label and do your research when selecting plants.

Before selecting and adding plants, it's vital to determine where and when the different exposures occur around your landscape. Step outside your home during different times of the day to see when and where sun and shade occur. The placement of buildings, walls, fences, trees, and shrubs will create shade. In the Northern Hemisphere, north-facing exposures are the shadiest, while those that face south receive the most sun (in the Southern Hemisphere, this is reversed). During the winter, shade will increase; in summer, it will recede somewhat. East- and west-facing exposures

**Different Sun Exposures around a House in the Northern Hemisphere**

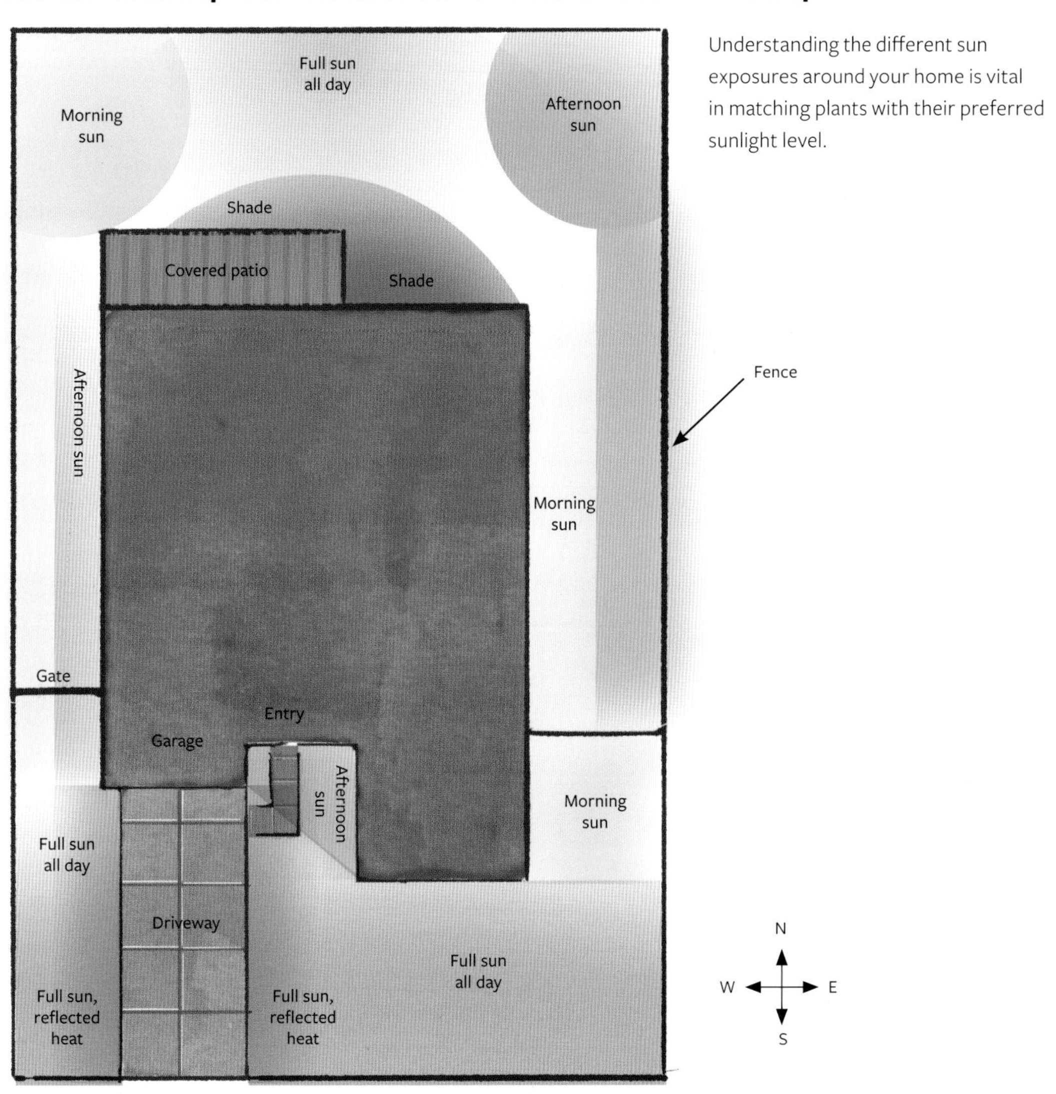


Understanding the different sun exposures around your home is vital in matching plants with their preferred sunlight level.

have periods of direct sun and shade during the day. The timing of sun also affects plants. For example, morning sun is milder than afternoon sun, which is hotter and more intense.

## Microclimates to Decrease Water Requirements

We can modify the climate outside our homes to create conditions that will help us conserve water. To do this, we create microclimates, which are small areas around your landscape with different climatic conditions than the rest of the site. Microclimates can be hotter and drier where maximum sun exposure occurs. Alternatively, we can create cooler and more humid microclimates than surrounding areas outside of your garden. Modifying the climate by lowering summer temperatures and increasing humidity decreases a plant's water needs.

**Microclimates** are created around a property by strategic plant placement to generate shade and lower temperatures to help decrease plants' water needs.

» The shade from trees moves with the sun during the day—east to west. Proper tree placement ensures that most areas will be covered in shade during the hottest part of the day—in the afternoon.

» Large shrubs planted in front of property walls and outside the home prevent the sun from heating up surfaces, keeping temperatures cool.

» Groundcovers throughout the landscape create an attractive swath of green, which use less water than a lawn and also help reduce surface temperatures.

### *Create Microclimates with Shade*

Shade is our greatest tool to help reduce hot summer temperatures and, therefore, reduce the water needs of plants. Plants grown in full sun generally need the most water. However, we can use strategies to minimize the amount of water plants need. First, if you have areas in the garden without shade, add native plants adapted to growing in full sun. You also can look at adding nonnative plants adapted to growing in climates similar to yours. Another option is to create shade for the area by adding a tree, which will allow you a wider range of plants to grow while providing protection from the sun's rays. Remember that the shadow from a tree moves with the sun throughout the day, so you should ensure that plants can handle brief periods of sunlight. By adding a tree, we have created a cooler and slightly more humid microclimate than the surrounding area. In addition to trees, fences, walls, and pergolas can be added to make shade.

To create adequate shade, where we plant a tree matters. My beautiful, large palo verde tree provides welcome afternoon shade to my west-facing front entry. Place trees where shade is needed most—near southern or western exposures where the sun is most intense. Have a large space that

This corner of the author's garden receives morning shade from the nearby tree but gets intense, reflected heat in the afternoon due to its location by the driveway and sidewalk. Artichoke agave (*Agave parryi* v. *truncata*), silver spurge (*Euphorbia rigida*), and angelita daisy (*Tetraneuris acaulis*) are all plants that can handle full sun and reflected heat.

An ironwood tree (*Olneya tesota*) provides welcome shade from the intense desert sun for aloe vera (*Aloe vera*), tropical milkweed (*Asclepias curassavica*), canna lilies (*Canna* spp.), and sago palm (*Cycas revoluta*) at the Desert Botanical Garden, in Phoenix, Arizona.

bakes in the sun most of the day? Consider adding a tree there. When planting a tree, keep it away from nearby pavement or walls where the roots can do damage. Tall shrubs may be a better option for those areas as their root systems are smaller.

### *Temporary Shade During a Heatwave*

During a heatwave, you may notice your plants are struggling—particularly those in full sun. Wilting leaves, yellowing, or burned foliage often point to heat stress. We can provide short-term relief for plants during these stressful periods by creating temporary shade. A valuable tool is shade cloth, which is a fabriclike material perforated by tiny holes that lets in small amounts of sunlight. When used to cover plants, shade cloth reduces the temperature while shielding the plant from direct sun. This useful material has different strengths based on how much sunlight it blocks. As a rule, shade cloth that blocks 50 percent of the sun's rays is useful for small trees, shrubs, and groundcovers, while 30 percent is better for cacti and other succulents.

Shade cloth protects an agave from reflected heat and sun in a desert garden during summer months.

## NOT ENOUGH ROOM FOR TREES?

Tall shrubs, such as the hop bush (*Dodonaea viscosa*) in this photo, can serve as a great companion or substitute for trees in narrow spaces where a tree won't fit. Use shrubs to create a tall hedge, provide shading near windows, and provide relief from the afternoon sun near a vegetable garden or patio. Planting shrubs against walls also can help reduce reflected heat from walls, lowering temperatures.

### *Reduce Drying Effects of Wind*

Windy days suck the moisture out of plants, which increases their water usage. Thankfully, we can do things to reduce the effects of wind in our gardens. One of the most effective strategies is to plant a hedge, consisting of shrubs planted in a row. The shrubs create a wind break, interrupting the wind's flow and dispersing it upward, away from plants growing lower down. The effectiveness of a hedge increases with the size of the shrubs used. Taller and broader shrubs block more wind (and sun) than smaller ones. Place the edge of hedges along the periphery of your property in relation to the direction the wind blows to provide protection. Consider opting for a naturally shaped shrub hedge rather than a formally pruned one as frequent pruning increases the water needs of plants.

California lilac (*Ceanothus* spp.) shrubs create a colorful focal point while helping to reduce the effects of wind and noise in this British Columbia, Canada, landscape.

A garden bed filled with young plants, including Pride of Barbados/ red bird-of-paradise (*Caesalpinia pulcherrima*) and lantana (*Lantana* spp.), which have enough room to grow to their mature size.

Drought-tolerant kangaroo paw (*Anigozanthos flavidus*) plants grouped with other low-water plants in a California garden.

## Group Plants That Share Similar Water and Soil Requirements

One of the easiest ways to maximize water efficiency in the garden is to place plants that need the same amount of water and sun, or shade, in the same section of your yard. To achieve this, we group plants that share similar water requirements. For example, put thirstier plants near each other to make frequent watering more manageable, and plants can share the water. When plants are watered (either by the rain or by you), the water doesn't stay within the root zone of a single plant. Instead, it spreads out in the soil, often beyond the root zone of the individual plant. Therefore, if you have plants with higher water needs grouped together, they can share the water as it spreads from plant to plant, which can help eliminate wasted water.

Plant drought-tolerant plants near each other as well, which allows you to match their watering needs more precisely rather than having higher water usage plants interspersed among them, which may end up wasting water or causing problems with excess water migrating toward plants with lower water needs. As discussed in chapter 2, most drought-tolerant plants don't like constantly moist soil. In regions with plentiful rainfall, you may have low-lying areas in your landscape where water tends to pool. Plants that need well-drained soil may struggle to survive in these wet conditions. However, mounding the soil helps it drain more quickly, making it appropriate for plants that need soil to dry out between watering cycles.

Alternatively, suppose you like the look of a mixture of different types of plants with varying water requirements. In that case, you can mix plants with differing water needs using drip irrigation. Admittedly, one of my favorite design styles has cacti and other succulents growing among shrubs and groundcovers. However, those two types of plants have very different watering needs. When combining plants with varying water needs, we need to be aware of spacing so that low-water

An Argentinian toothpick cactus (*Stetsonia coryne*) adds spiky texture contrast in front of Bells of Fire™ tecoma (*Tecoma stans* 'Bells of Fire'). Shrubs are planted several feet away to prevent problems with overwatering with the cactus.

plants don't get water from those watered more frequently. A drip irrigation system can be set up so there are two separate irrigation lines (valves)—one for the succulents and the other for the shrubs/groundcovers. Different watering schedules can be implemented by providing separate water lines to match the water requirements pairing of distinct plant types.

A side yard with overplanted and overpruned purple-flowering shrubs. The plants don't have room to grow to mature size, requiring frequent pruning and detracting from their appearance—not to mention seldom flowering and increasing water usage.

Newly planted dwarf morning glory (*Evolvulus glomeratus*) spaced to cover the bare soil as they grow to their mature width.

## Allow Plants to Grow to Natural Shape and Size

Drive down any neighborhood street, and you will likely notice several yards with compact, neatly clipped trees, shrubs, and groundcovers. This landscape style is very popular, yet it is hiding a secret. Plants frequently pruned into smaller versions of their mature size often need more water than if allowed to grow larger. Less is more concerning pruning frequency (see chapter 7, page 113). The truth is that allowing plants to reach their full size conserves water by creating shade over the root zone, which prevents soil from drying out too quickly. When the bare ground around plants is exposed to sunlight, moisture will evaporate more rapidly.

Unfortunately, many homeowners and even some landscape designers and architects tend to overplant a new landscape area to create a filled-in appearance. While this may look great initially, problems soon ensue once plants grow and crowd into each other. Do you have too many plants in your existing landscape? Remove excess plants, leaving enough room in between for the remaining plants to grow to their preferred size. When designing a garden, ensure there is enough room for plants to grow. By avoiding overplanting, you'll save money, maintenance, and water, and your landscape will look better!

### Minimize the Distance Between Plants

In a dry climate region, you'll often see plants with empty space covered in organic or rock mulch between plants. This spacing is due to arid conditions with less rainfall and less competition for water availability for the roots of each plant. If you live in a dry or semi-arid climate, this is often the preferred method of plant placement. Therefore, allowing enough room for plants to grow to their natural shape and size is essential to take up as much space as possible and shade the soil in their area. If you plant more densely, that is an option, but it will use more water.

In places with more plentiful rainfall, you don't need to have significant gaps between individual plants unless you want to. In fact, by minimizing the space between plants, we can create more shading of the soil, which helps prevent weeds and the soil from drying out too quickly. Closer spacing can help create cooler microclimates throughout the garden. Additionally, a plant's roots also contribute to the ability of the soil to hold onto moisture, so more plants nearby allow the soil to retain more water for a larger area. So, how do we create this ideal spacing without letting plants overcrowd each other? Take the width a plant will reach at maturity and use that measurement for how far to space adjoining plants—it's okay for plants to touch each other; we just don't want them to grow into each other and hamper growth. When practicing this method, it's important not to pair an overly aggressive plant with one with a slow or moderate growth rate, which can lead to one growing into and impeding the growth of the slower plant.

## IMPATIENTLY WAITING FOR NEW PLANTS TO GROW?

New plants can initially seem rather small in a garden with lots of bare space around them. While it's tempting to compensate by adding more plants that will eventually outgrow their space in the future, fill those temporary bare spots with annuals or short-lived, drought-tolerant groundcovers while waiting for your new plants to grow larger.

A landscape planted with Caribbean agave (*Agave angustifolia*), chocolate flower (*Berlandiera lyrata*), and bush lantana (*Lantana camara*) spaced closely together to help shade the soil and prevent it from drying out quickly at the Desert Botanical Garden in Phoenix, Arizona.

CHAPTER 4

# Building Drought-Resistant Soil

**When ensuring we are making** the most of water in our gardens, we must turn our attention to the ground beneath our feet. Plants absorb water and oxygen from the soil, and the type and condition of our soil play a crucial role in how much water is available for our plants. It's helpful to think of soil as a "bank" where water is stored temporarily. When we water plants, water fills the spaces between the soil particles, temporarily forcing out the oxygen in those spaces. Healthy soil allows water to permeate deeply into the soil instead of running off, which promotes deep root growth for plants and is essential to ensuring they can access water and go longer periods between watering. However, if your soil is in poor condition, it cannot do its job well. Problems include soil staying wet too long, water running off without permeating to plant roots, drying out too quickly, and requiring more frequent watering. If we don't have the correct soil type that plants need, they will suffer, and they may use water less efficiently. Thankfully, you can implement strategies to create healthy soil for plants to make the most of the water they receive.

Colorful, waterwise plants such as firesticks (*Euphorbia tirucalli* 'Sticks on Fire') and Coulter bush (*Hymenolepis crithmifolia*) grow best in well-drained soil at the Getty Center in Los Angeles, California.

The author digging holes in the sandy soil of a Savannah, Georgia, garden with an auger, which makes digging many holes less strenuous than using a shovel. For areas with hard-packed soil, soak planting sites with water a day or two ahead of time, which will make digging out soil easier.

Know what texture your soil is before planting so you can determine its water-holding capabilities and add compost, if necessary.

## Clay, Loam, and Sandy Soils

Before we delve into improving your soil's water-holding capabilities, it's important to understand the different soil types and how they retain water differently. Soil types refer to the kind of "texture" a soil has, which refers to the size of individual soil particles ranging from clay, loam, and sand. While soil may not be an exciting topic to discuss, it is vital to selecting the right plants and making any adjustments with soil amendments to modify its water-holding capacity. Our plant choices can depend on the type of soil we have to ensure their success. If we grow plants that are native or adapted to the kind of soil we have, we may not have to amend the soil to improve it. However, if you decide to incorporate plants that prefer a different type of soil than you have, we can add amendments to the soil to help it retain water or drain more easily. Let's begin with the three main soil types.

### Clay Soil

Clay soils have individual soil particles that are very small and flat, which create a smooth, claylike texture when wet. A wet lump of clay will feel heavy in your hand and will become very hard if allowed to dry. When watered (either by rain or irrigation), clay soil takes a long time to absorb the water, and if water is applied too quickly, it will run off without permeating into the ground. Once wetted, clay soil can take several days to dry out. Most drought-tolerant plants do best in well-drained soil and can struggle in heavy, clay soil that takes a long time to dry out, which creates a lack of oxygen available for the roots of plants.

### Sandy Soil

Soils with sandy textures are on the opposite end of the spectrum from clay soils. They have large soil particles and large spaces between them where water and oxygen reside. When wetted and rubbed between your fingers, sandy soils have a granular, sandy

## Soil Texture Jar Test

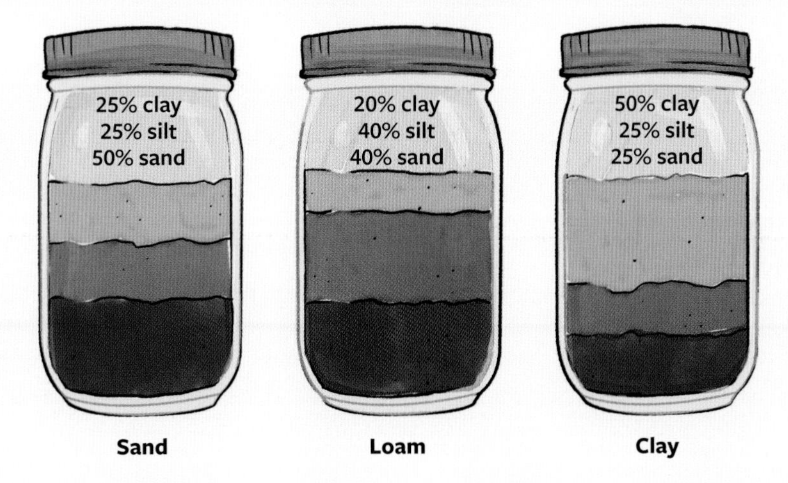


**Most soils are a varying mixture of clay, sand, and silt. To determine the makeup of your soil:**

1. Take a mason jar with straight sides and fill it halfway with soil from your garden, removing rocks, roots, and any clods.
2. Add water, stopping an inch from the top.
3. Add a tablespoon of powdered dishwasher detergent (or borax) to the soil/water mixture to help the soil layers separate into defined layers.
4. Shake up the contents thoroughly.
5. Wait one minute and mark the first bottom layer where the sand layer has collected.
6. After 2 hours, mark where the top of the next layer is, which is silt.
7. In 24 hours, mark the top of the upper layer, which is the clay layer.

texture. Water moves quickly into and through sandy soil due to its larger soil particles and permeates rapidly into the root zone of plants. However, sandy soil will dry out quickly as water moves downward away from plant roots or evaporates into the air. Plants grown in sandy soils need more frequent irrigation than those in clay or loam soil.

## Loamy Soils

If you are fortunate enough to have soil described as "loam," consider yourself lucky—this is the ideal soil for growing plants and maximizing the soil's water storage capacity. Loamy soils consist of a mixture of clay, sand, and another soil type called silt. Silt has medium-sized soil particles from eroded rock and has a smooth texture, similar to dry flour. A loam soil consists generally of equal amounts of sand and silt and slightly less clay. A loamy soil holds the right amount of water, not becoming too wet or dry because there are medium-sized spaces between the soil particles.

The ideal soil is shown in the middle jar, which is loam. The jar on the left is composed of mostly sand while the one on the far right has more clay content. Soils with higher sand or clay content will benefit from the addition of compost to create the right balance of water-holding capacity.

In addition to the soil texture jar test, another way to get a general idea of your predominant soil type is by the "feel method." To perform this test, thoroughly moisten a small amount of soil in the palm of your hand and rub it with your finger. If the texture feels gritty, your soil has significant amounts of sand present, while smooth soil texture indicates higher clay content. A loam soil will have a more uniform texture without sand or clay predominating (for more information on soil texture, see the Resources section).

## The Magic Ingredient for Water-Efficient Soils

Healthy, loamy soil will absorb water like a sponge that holds onto water for plants without becoming overly wet. However, suppose your soil test reveals mainly sand or clay soil. In that case, amending your soil may be necessary to modify its water-holding capacity to ensure healthy plants. Compost works wonders to help transform clay and sandy soils by changing how they retain water. The organic matter that compost adds helps the soil form a better structure and increases the amount of spaces between the soil particles in which to store water. The addition of compost allows the soil to hold onto the right amount of moisture without drying out too fast or becoming waterlogged. In other words, incorporating compost into clay soil allows excess water to drain more quickly, while with sandy soil, compost will let it hold onto water longer.

For people who live in arid or semi-arid regions and want to grow native or arid-adapted plants, adding compost is usually unnecessary and may cause problems. These plants may struggle if grown in heavily amended soil with compost added, which can hold onto too much water—this is especially true for cacti and succulents, but it also can affect other plants. For these areas, compost is best used for plants that aren't native and don't come from arid regions.

Gypsum is a frequent suggestion as an amendment to help break up heavy clay soils, but studies have shown that gypsum has little to no long-term effect in improving the drainage of clay soil. Additionally, gypsum can decrease the fertility of the soil. Furthermore, its effects only last a few months and don't last as long as compost; see Resources section for additional information.

### How to Test for Soil Drainage

It's crucial to know how well your soil drains so that you can select appropriate plants or make modifications to the soil. Thankfully, there is an easy way to test how well the soil in your garden drains. Dig a hole about 1 foot (30 cm) deep and wide—you also can use a future planting hole for the test. Fill the hole to the top with water and let it drain. Then, fill the hole with water again. Water will drain within three hours in sandy soils but

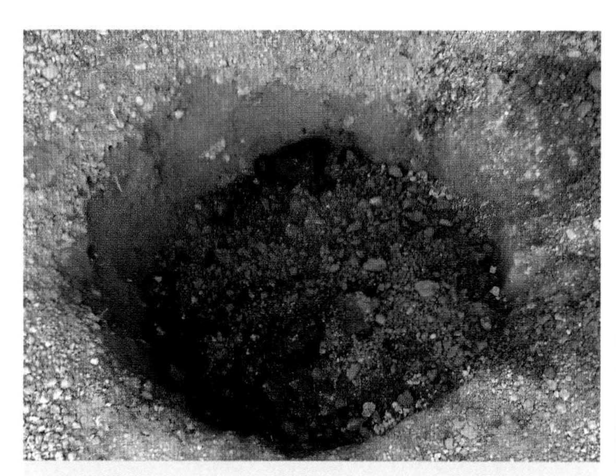

Compost is added to a hole made in clay soil before planting to help improve drainage.

Fill a hole with water the day before planting to check for drainage characteristics.

## WHAT NOT TO ADD TO IMPROVE SOIL DRAINAGE

If you have heavy clay soil, you may think that adding sand is an easier option than using compost. However, if you add sand to clay soil, it won't help it to drain better. You can make it worse as the sand creates a gluey to concrete-like texture as the clay particles fill the spaces between the sand. There is also a danger of this occurring if you add clay to soil with a sandy texture. Use compost to improve the water-holding capabilities of clay and sandy soils.

A 2-inch (5 cm) layer of compost is added annually to this garden bed in northern California to improve soil texture and water-holding capacity so that soil doesn't stay too moist or too dry.

may take twelve hours or longer if it has more clay content. If your soil drains slowly, incorporate compost to help improve your soil to ensure a healthier ratio of water retained in the soil if using plants not native to your region.

### How to Apply Compost to Improve Drought Resilience of Soil

If you are planting a new garden area, apply a 2- to 4-inch (5–10 cm) layer of compost and incorporate it into existing clay (or sandy) soil 6 to 12 inches (15–30 cm) in depth—this process is called tilling the soil. You can do this with a shovel or a rototiller. It is important to note that tilling should only be done once for new garden spaces and not repeated as it disrupts the healthy soil texture and brings weed seeds up, which will germinate. For individual plant holes, you can fill the hole one-third full of compost and mix it with the existing soil. The compost will intermix with the existing soil particles to create healthy spaces for the water to flow into.

In existing garden areas, topdress around plants by adding a 2-inch (5 cm) layer of compost around plants. The compost will gradually improve the texture and quality of soil as it breaks down, while simultaneously modifying the water-holding capacity. Apply compost annually around plants in spring or fall. You also can use fallen leaves to create a compost layer. So, instead of blowing or raking your fall leaves away, use leaf litter around plants as a substitute or mix with compost around plants. Make sure to shred larger leaves into smaller pieces to allow them to breakdown more easily into the soil (see chapter 7, page 111).

If you have areas of compacted soil, till the soil, incorporating compost. Topdress with 3 to 4 inches (8–10 cm) of compost the following spring or fall. Do this every year, which will help to decrease soil compaction. To help avoid further compaction, place stepping stones in areas where you need to walk to prevent any more compaction from occurring.

CHAPTER 5

# Rain and How to Capture It—Free Water!

**Water is a big deal in the garden**—without it, our plants won't grow. All the water we use to irrigate our garden comes down from the sky as precipitation in the form of rain, snow, and hail that moves to lakes and rivers and infiltrates down into aquifers. The amount of fresh water available for our use is minimal—only 0.5 percent of all the water on Earth. Household water that goes toward outdoor use ranges from 30 percent per household in wetter climates to 70 percent in drier locales. That is a lot of water! But what if we could decrease the amount of water we use from the tap by utilizing more of the rain that falls within our gardens instead of letting it drain away? The benefits are many, from a reduced water bill to a more resilient garden that isn't as dependent on outside water sources. Moving forward, let's reexamine how we use water resources within our gardens by exploring how rain can play a larger role in the garden by looking at how you can capture and use it.

A flowering yucca enjoys a welcome rain.

For climates with plentiful rainfall, rain is frequently the primary water source for plants, and supplemental irrigation may not play a large part. But what about during dry seasons or in times of drought? How often have you hauled out your hose to water plants or turned on your irrigation system? For people who live in drier climates, reliance on regular supplemental irrigation for plants throughout most of the year is standard in most gardens. So much reliance on supplemental irrigation is especially concerning when excess precipitation is allowed to drain away from the garden and run down the street. However, we can change our view of rain and implement strategies that will enable us to use rain as a major resource for our watering needs.

The incidence of drought is increasing worldwide, with many regions experiencing long-term dryness for decades. If you live in an area affected by drought, trying to figure out how to provide water to your plants can be overwhelming. Whether you experience drought conditions or want to minimize input from outside water sources, we can make significant progress by using rain that falls on and around your home. Imagine retaining much of the precipitation that falls within the boundaries of your property to water your plants rather than needing to provide extra water for plants to grow. So, let's create more waterwise landscapes by strategically capturing rain in the landscape areas around our homes by exploring rainwater harvesting.

A shrub showing signs of salt buildup in the soil with browning leaf tips.

### FREE FERTILIZER

Did you know that rain helps to fertilize your plants? Rain captures nitrogen from the atmosphere and brings it down to the plants. That is why plants look greener after rain has fallen. Nitrogen is an essential nutrient for plants, and rain brings it to you for free. When you use captured rainwater to irrigate your plants, they also get this benefit.

## Why Rainwater Is Best for Plants

Rainwater is of higher quality than the water from your hose. To understand why water from rain is better than tap water, we need to discuss salts naturally present in water. All water contains dissolved salts; some sources have more salts than others. The water that comes out of your tap contains a higher percentage of dissolved salts than rainwater. So, why do salts in the water you use to irrigate your plants matter? Because salts can accumulate in the soil around the root zone of plants. Not surprisingly, plants can struggle to survive in salty soils (see chapter 6, page 88). The salt level in soil builds up over time, particularly in arid regions where rainfall amounts are lower. Rainwater contains very low amounts of salt, which makes it a superior source for watering plants. It can also help reduce the salts already present in the soil, helping to flush them away from the root zone of plants.

## Maximizing the Benefits of Rain in Your Garden

To begin our rain-harvesting journey, we need to look at how we view and use rain. Many gardeners appreciate the rain that falls, knowing it provides a welcome drink for their plants. Any excess water that doesn't soak into the ground right away is then routed away toward the street, only to disappear down the gutter. Therefore, any benefit from rain is significantly reduced because it flows away quickly before it can soak into the soil. In the landscape areas around homes, properties are often graded at a slight slope to channel water away from your property and into the street. Rain gutters around your home's roof and dry creek beds or swales lined with stone are frequently used to direct water from the garden into the street. These strategies route water away from the home to help mitigate flooding risks. However, in many cases, it's possible to retain a significant amount of rainwater without the risk of flooding.

Imagine retaining much of the precipitation that falls within the boundaries of your property to water your plants. There are many benefits to rainwater harvesting—first, it's free! Water bills can be expensive and will likely increase as rivers, lakes, and groundwater supplies shrink. Additionally, with decreasing water supplies and increasing outdoor water restrictions in regions affected by drought, making the most of rain that falls in your garden allows you to be more independent of supplemental sources of water. With the increasing frequency of drought and flooding events, capturing rain strategically in the landscape areas around your home can create a more resilient garden.

In drier regions, rainwater harvesting makes particular sense. However, if you live in an area with plentiful rainfall, why should you consider implementing strategies to route and store the water from rain throughout your garden? Harvesting rainwater allows rainfall time to soak into the soil; it percolates into the ground to help replenish groundwater levels. When water flows down the street, it picks up pollutants. Rainwater harvesting can help to reduce polluted water from reaching rivers, lakes, and oceans while helping to reduce flooding risks.

The plants in this Texas garden receive most of their water requirements from rain.

As you begin your rainwater-harvesting journey, start by collecting rain as it runs off the roof into buckets as you work toward additional harvesting strategies.

## How Much Rain Can You Harvest around Your Home?

To put into perspective how much rainwater your property receives, an inch of rain that falls on a square foot of space equals 0.62 gallons (2.3 L). That is a lot! Imagine capturing and using the rain that falls on the roof of your house, driveway, patio, and the ground around your home. In my dry, desert region, where annual rainfall only averages 8 inches (20 cm), that amounts to almost 5 gallons (19 L) of water per square foot of my property. For a property that is a quarter of an acre, that amounts to 54,450 gallons (206,116 L) per year in the low desert where I live. If you live in a climate with less aridity, you'll receive even *more* free water that you can use.

# Active and Passive Water Harvesting

Let's explore the two methods you can use to make the most of the rain you receive instead of sending it down the drain. Simply put, harvesting rainwater is the practice of keeping the water for future use by storing it or channeling it to areas around your home. There are two main ways we do this. First is active rainwater harvesting, which entails storing water in containers such as rain barrels, water tanks, or underground cisterns for later use. The second type is called passive rainwater harvesting, which incorporates strategies that channel water into the soil directly. Whether you use just one type of harvesting method or both, you can reduce your dependence on outside water sources.

## Active Water Harvesting: Collecting and Storing Water

If you've ever put out a bucket to capture rain from the roof or to collect water while waiting for the shower to warm up, you've actively harvested water for plants on a small scale. Active water harvesting, or rainwater storage, allows us to capture and store rainwater until we need to use it to water our plants later. This type of water harvesting is beneficial for regions with rainy seasons interspersed with months when little to no rain falls. Stored rainwater can provide supplemental water for plants in dry periods. Let's explore ways to harvest water by looking at three larger types of vessels for storing rainwater—rain barrels, water tanks, and cisterns. Typically, the rain that falls on your roof is routed toward one of these types of vessels, where it is kept until needed.

A rain barrel captures rain from the roof of a garden shed for use in a Minnesota garden.

## CHECK LOCAL ORDINANCES BEFORE INSTALLING ACTIVE WATER-HARVESTING SYSTEMS

It's vital to research rules and possible restrictions for using water catchment vessels such as rain barrels, water tanks, and cisterns. If they are allowed, there will likely be guidelines to follow and permits required. Check with your municipality or neighborhood rules before getting started. Additionally, some cities offer rebates for residents who implement water-harvesting strategies.

Rain runs off a roof, with most of it flowing into the street instead of being harvested and used within the landscape.

***Determining Rain Capture from Your Roof***

When planning out your active rainwater-harvesting strategy, it's helpful to know how much rainwater you can capture from your roof so we can determine what size vessel will work best. To do this, we need to figure out the square footage of your house—specifically, how much of your house is covered by the roof. If you already know this—great! However, garages aren't always included within square footage calculations for real estate purposes, so you may need to add in the square footage of your garage. If you don't know your square footage, it's easy to figure it out. Measure the length and width of your house, which will match the general measurement of your roof. Then, we need to use those measurements to get the area (square footage) of your roof.

Once you know how much rain falls on your roof, it's time to decide how much rain you want to harvest and store for later use. If you live in an area where rainfall is plentiful throughout most of the year, a rain barrel may be a good choice for minimal supplemental watering. Average sizes of rain barrels for residential use can vary from 50 to 90 gallons (189–341 L). In more arid regions or those with prolonged dry seasons, an above-ground water tank or cistern may be a better choice to hold more water.

## Gutters, Rain Chains, and Scuppers

Rooftop rain can be routed down and away from the house in several ways. The most effective method for capturing rain from the roof is rain gutters, which are narrow troughs installed around the roof line to catch water as it sheens off the roof. Water from the gutters is routed toward a downspout, which brings the water down to ground level, releasing it. Active water-harvesting systems such as rain barrels and water tanks can connect to

## Rooftop Water Harvesting

Calculating the amount of rain you can capture from your roof allows you to determine which water-harvesting strategies will enable you to harvest as much rainfall as possible.

An inch of rain equals 0.623 gallons (2.4 L) per square foot, and about 90 percent of the rain that falls on your roof can be captured. Using these factors, we come up with the following calculation:

***Rooftop Rainwater-Harvesting Formula:***
Square footage of your house × inches of annual rainfall × 0.623 gallons (2.4 L) × 0.90 = the amount of water you can harvest from your roof.

For example, a roof that is 40 feet wide and 60 feet long (12 × 18 m) has 2,400 square feet (223 sq m) of area. Annual rainfall is 12 inches (30 cm). Using the equation above:

2,400 (roof area) (223 sq m) × 12 (inches rainfall) (30 cm) × 0.623 gallons (2.4 L) × 0.90 = 16,148 gallons (61,127 L) of potential rainwater you can collect annually for later use from your roof.

A house in Tucson, Arizona, without rain gutters uses plants under the roofline to absorb water from the roof, decreasing runoff.

a rain gutter system. Rain chains also can connect to the gutters and are a more decorative alternative to downspouts and help channel water down from the roof. However, they cannot handle large amounts of water from heavy rainfall as efficiently as a downspout can. For flat-roofed buildings, rain from the roof flows through scuppers, which are openings high up on the sides where water spills down to the ground. Finally, in dry climate regions, many roofs don't have rain channeling systems, such as gutters, due to lower amounts of rain. If you have no gutter or scupper system, consider adding plants around your roof line to help absorb the rain as it drips from the roof, or have a gutter system installed.

### *Rain Barrels*

Rain barrels are the least expensive and easiest to install water-storing devices. Locate rain barrels on a raised, solid 6-inch (15 cm) base next to the house. Gutters channel the rain from the roof via a downspout into the rain barrel. The top of the barrel is covered, and water is filtered through a screen to remove debris and prevent mosquitos from accessing the water. Toward the top of the rain barrel is an overflow valve on the side, which is a short pipe that allows excess water to flow out once it gets near the top of the barrel to prevent it from overflowing. At the base of the barrel is a spigot, which allows you to connect to a drip irrigation system, to a hose, or to fill a watering can to bring water where needed. Water flow through the hose is gravity-fed and must be used to water plants in areas lower than the barrel. In homes with small children, rain barrels should be fastened to secure support to prevent toppling over and have a secure lid. Online tutorials or an irrigation professional can help you set up your rain barrel.

Rain barrels connect together to maximize the amount of rainfall captured from a garage in Seattle, Washington.

## WINTER PREP FOR RAIN BARRELS AND TANKS

In cold winter climates, empty above-ground water storage vessels such as rain barrels and water tanks before freezing temperatures occur to prevent cracking because water expands once it freezes. Alternatively, you can insulate your water tank to prevent the water from freezing. Check with local experts on the best way to winterize your active water-harvesting systems.

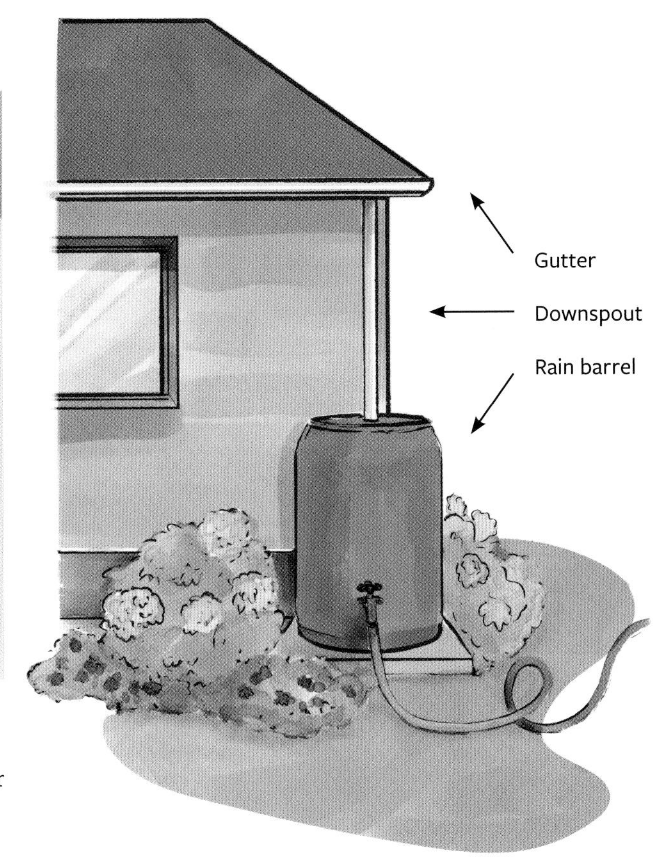


**Rain Barrel Set-Up**

A rain gutter channels water from the roof through a downspout into a rain barrel, where it can be stored for later use.

While rain harvested from your roof is okay to use for ornamental plants, it may not be suitable for watering edible plants since it can contain possible contaminants from the roofing material. If using rain barrel water for fruit and vegetables, make sure to keep water off the foliage and focus water at the soil level. Avoid using the first tank of water after a prolonged dry spell on edible crops because it may have higher levels of contaminants. Instead, use it on ornamental plants. For more information on using water from a rain barrel for edible plants, look at the Resource section.

### *Above-Ground Water Tanks*

Water tanks (also called above-ground cisterns) work similarly to rain barrels but on a much larger scale. They are typically located next to your house and allow you to store much more water—anywhere from 100 to over 1,000 gallons (378–3,785 L)—and are installed by irrigation professionals. Water is routed through a gutter system from the roof to the tank for future use. An overflow valve toward the top of the water tank directs excess water away from the house to other parts of the garden, connects to another tank, or flows down the street. As with rain barrels, water from water tanks is distributed to plants via connection to drip irrigation, hose, or watering can. An electric pump will enable water to reach different areas of the garden. Multiple water tanks can connect to increase storage capacity, or excess water can be routed to underground cisterns, if present, via pipes once the water tank has reached its maximum capacity.

### *Underground Cisterns*

Cisterns have the greatest capacity for storing water that is routed from the roof and, potentially, from other parts of the garden. Many cisterns are located underground in large holes and connected to pipes that direct the water from the roof and other areas. Cisterns can operate with rain barrels and above-ground water tanks, storing excess water from those receptacles once they have reached

Rain chains direct rain from the gutters into underground pipes that lead to a cistern in the front yard.

capacity. The size of underground cisterns varies but averages 5,000 gallons (18,927 L) for residential use. A pump is required to access water from a cistern to allow for connection to your irrigation system. Cisterns are the most expensive water storage option due to the digging required, the size of the tank, and the pump system needed to allow you to use the water. However, they will enable you to store larger amounts of water and rely much less on other sources for watering your plants.

To determine what type and size of water-storage vessel will work best for you, begin with the rooftop calculation of water from your roof—that is the maximum size you will need (see chapter 5, page 69). You may be fine with a smaller cistern if rain in your area doesn't fall within a short period of time, and you will use water in between rain events. In general, if you experience long periods without rain, you may want to opt for a storage vessel that can hold more water during the rainy season that you can use later when the weather begins to dry out. Rain-storing vessels come with an overflow valve, so excess water will be routed away toward other parts of the garden.

### *Greywater Systems (Recycled Household Water)*

A different type of active water harvesting doesn't involve collecting rainwater but reusing household water. Greywater harvesting is the practice of using water that drains from bathroom sinks, washing machines, and your shower to water

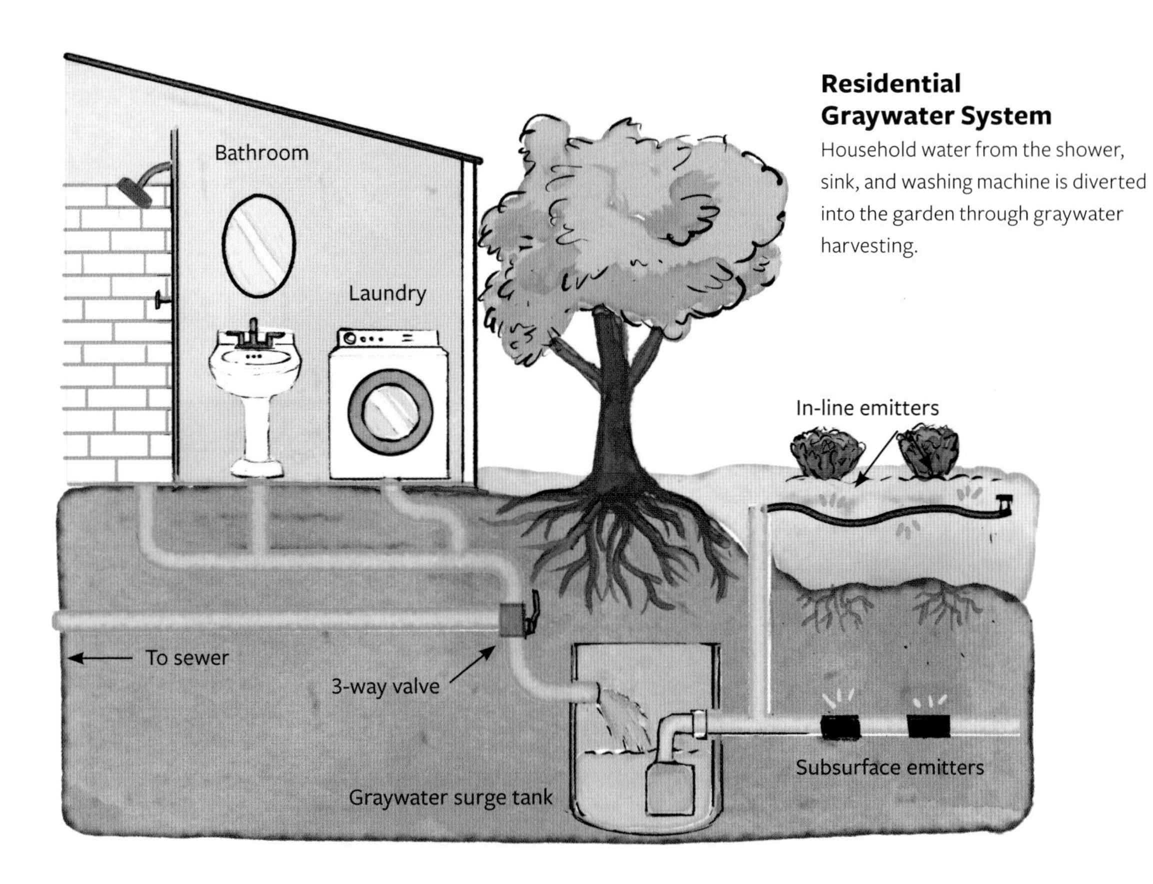


**Residential Graywater System**
Household water from the shower, sink, and washing machine is diverted into the garden through graywater harvesting.

plants, instead of sending it down the drain. A basic form of greywater harvesting is collecting water from the shower in a bucket while waiting for the water to warm up. Greywater sources are suitable for ornamental plants. This type of household water is called grey because detergents, bacteria, and other materials are contained within it, but it is still relatively clean. Other household water sources—the toilet, dishwasher, and kitchen sink—are considered black water and aren't safe for reuse in the garden. A licensed plumber or contractor, experienced with greywater systems, modifies the current plumbing in your home to allow for the capture of the water for use in the garden. Greywater is then used to water plants via a hose or connected to a drip irrigation system. Greywater must be used within a day of its collection to prevent odors.

While greywater is a useful way to maximize your water use, there are some restrictions to keep in mind. Some communities require permits before installing a greywater system. Another essential factor is that greywater isn't suitable for plants that will be eaten, such as in vegetable gardens. Greywater isn't suitable for use on root vegetables or edible crops that come in contact with the ground. If using greywater in your drip irrigation system, it can plug up emitters, so be sure to periodically flush out your irrigation system and replace emitters as needed (see chapter 6, page 100). Additionally, the type of shampoo, toiletries, and laundry detergent you use can affect your plants. Using bleach and borax can harm plants, so it's important to have a diverter valve to direct water directly to the sewer instead of outdoors when you use those products. See the Resource section for more information on what products to avoid using with greywater systems.

## Passive Water Harvesting

When it rains, plants receive some water, but if your homesite is graded to channel water away from your house, gravity always moves water flow away from the land around your home and toward the street and storm sewer. Consequently, plant roots receive only a fraction of the fallen rain. Passive water harvesting involves strategies to both slow down and capture runoff from precipitation, allowing it to water plants and infiltrate into the soil. This type of rainwater harvesting uses soil to shape and contour the landscape to channel and slow the runoff from rain. By slowing down water's progress, water can soak deeper into the ground and flow toward plants using shallow depressions called basins or swales. Shaping the soil to capture water is an ancient practice still utilized today and is very effective at maximizing the reach of rainwater for your plants. It's helpful to imagine your garden acting as a living sponge, absorbing rain to provide water for your plants. The soil in your garden is the best "water bank" to store water for immediate and later use.

Choosing where to create channels and basins begins with a trip outdoors during (or after) a rainstorm. Observe where rain falls and flows during a rain event around your home. Note areas where rain runoff occurs, such as from a downspout or the roof. Does water tend to pool in certain areas? Are there drier areas of your garden where rain seldom accumulates? Knowing how rainwater moves in your landscape is helpful when deciding where to add contouring with soil to direct water toward plants and away from buildings—research rules regarding rainwater harvesting where you live. In general, implement passive rain harvesting strategies at least 10 feet (3 m) from structures (or farther away if there is a basement) to prevent flooding.

A rock-lined channel directs rainwater away from walls and toward plants.

Here are several examples of how to use active harvesting strategies in your landscape:

### *Channeling Water through the Landscape*

The decision of where to move water through the landscape does involve some strategy. When you observe how water moves through your landscape during rain events, note where natural water channels occur—these are often an excellent place to start, as water is already flowing along these routes. To help direct water to specific areas in the garden, create 12-inch (30 cm) deep channels (trenches) lined with rock to move water through the landscape. The width of channels is typically 1 to 2 feet (30–61 cm). These trenches must move away from buildings to prevent flooding as gravity moves water along the trenches. All passive water-harvesting strategies should route water away from structures and allow overflow to drain toward other parts of the garden and, perhaps, ultimately, to the street.

### *Berms (Mounds)*

Creating contours within your landscape is how we direct water as it flows away from the house and other hard surfaces. When we create berms, we take soil and create a mound with gently sloping sides. The mounds create height throughout the landscape, and water moves around them. Place plants that require well-drained soil along the sides of the mound where water is sure to drain away quickly. The height of mounds should range from 12 to 20 inches (30–51 cm) at the peak and slope downward. The height of the mound will settle a couple of inches as the soil slowly settles over time. You can order topsoil to create the mounds or use excavated soil from your property that was removed to make a swale or channels. Avoid using cheap soil or fill dirt for building up a mound, which is often filled with weed seed and is of poor quality.

Runoff from a house is directed toward plants by contouring the landscape with berms.

### *Basins, Catchment Basins, and Swales*

As we channel rainwater, we need to create shallow basins for water to collect and slowly infiltrate into the soil. A simple form of a water catchment basin is making a planting well (basin) for a newly planted tree or shrub to concentrate water at the root zone, which allows water to infiltrate deeper into the soil instead of running off quickly. These basins allow water to remain available for any nearby plants and will enable it to penetrate deep into the ground, eventually reaching groundwater levels. Larger basins (often called catchment basins and swales) can differ in size, but an average depth is generally 6 to 12 inches (15–30 cm) deep. The width can range from 16 to 18 inches (41–46 cm), while the length can vary from short to very long. The sides of basins should have a gentle upward slope on the sides—the side that faces away from buildings should be slightly lower than the other sides to allow excess rainwater to flow out. Line basins with a layer of landscape rock using plants

At the Scottsdale Xeriscape Garden in Arizona, a concrete catchment basin is designed to fill with water channeled from the nearby building, slowing its progress through the landscape. Once the basin fills, excess water runs off along a channel to water nearby sideoats grama grasses (*Bouteloua curtipendula*).

### Active and Passive Water Harvesting

Rooftop water is directed to a water tank. Excess water flows through the overflow valve into a series of connected swales, channels, and a rain garden.

located around the side—it's okay not to use plants. If the soil doesn't drain well, adding plants will help water drain faster, preventing problems with mosquitos. Basins should be placed at least 10 feet (3 m) from buildings to avoid the risk of flooding and several feet away from trees to avoid problems with roots and overwatering your tree. Contact local utilities to ensure there are no cable, electric, gas, or water lines where you are digging.

Evaluate the effectiveness of your swale when it rains to be sure no modifications are needed. Multiple swales can connect by above-ground channels or underground pipes.

## REDUCING FLOOD RISKS AND POLLUTANTS IN RUNOFF

Passive rainwater harvesting can help your neighborhood and community by helping to reduce flooding risks from rain runoff entering the street. When rainwater runs down the road, it picks up pollutants on their route toward the storm sewer. The polluted water eventually makes its way to streams and rivers. By rerouting and retaining rainwater, we provide a free source of water for plants and allow rain to infiltrate into the soil, eventually contributing to groundwater.

### *Rain Gardens*

A popular way to capture rainwater around homes is rain gardens, a type of swale where plants are growing within the shallow depression of the basin. Rain gardens retain more water than a lawn, which contributes to plants reducing runoff into the street or flooding from rain events. Rain gardens can be personalized for your needs regarding size and can be as small as 2 feet (61 cm) in width or much larger. Trees, shrubs, perennials, groundcovers, and grasses are suitable for a rain garden. Select plants known to have deep root systems help open the soil to allow water to infiltrate downward. If your rain garden is often wet, select plants that can handle wet soil conditions over a long time.

Conversely, if you live in a region with long periods of dryness between rain events, choose native or drought-tolerant plants adapted to your climate. Rain gardens shouldn't be placed in poorly drained areas as the water in a rain garden needs to drain within twenty-four hours to prevent mosquito breeding. See how to create a rain garden in chapter 11, page 169.

### *Harvesting Water from the Driveway, Sidewalk, and Patio*

A great source of rainwater comes from the impervious surfaces around our homes, such as asphalt or concrete, where runoff typically makes its way to the street. Harvesting water runoff from your driveway or roof can be as simple as adding plants next to it, which help the water to soak into the soil as plant roots help to open the soil for water to infiltrate. Alternatively, you can increase the amount of water captured from the driveway by creating a shallow channel (trench) along the side of your driveway with a basin (or rain garden) at the end where water can collect. Line the channel with landscape rock, to prevent erosion. As water collects in the basin, it waters the existing or new plants planted around it. A channel alongside a sidewalk or patio effectively captures water by slowing it down so it can water plants nearby. Channels also can direct water away from hard surfaces toward planted areas several feet away.

The gentle swale of a rain garden in Michigan with a terrace wall that helps to slow down runoff to the street, allowing water to infiltrate into the soil.

Rain from a driveway runs into a swale where plants enjoy an extra drink of water and where the swale reduces runoff into the street.

### *Terraces*

Keeping water on sites with steep slopes is difficult without water-harvesting strategies in place. Rainwater can seldom permeate the soil in sloped areas as gravity pushes it quickly off-site. This rapid removal of water often results in erosion of the soil. To help prevent erosion and allow rain to permeate the soil, we need to break up the slope into flat steps called terraces. Flat, raised areas are built into sloped areas and are edged with wood, brick, concrete blocks, or rock along the front and sides. Each level of a terrace serves to interrupt the slope of a given area and the flat area, allowing rainwater to slow and percolate into the soil before flowing down to the next level. Terraces can consist of multiple levels, depending on the size and steepness of the slope. For areas with steep slopes requiring terraces taller than 18 inches (46 cm), consult a landscape professional familiar with creating terraces. There are many types of materials to consider for terraces. Treated wood is a popular option because it is relatively inexpensive compared to other materials and is easier to work with. However, walls made from heavier materials such as brick, concrete blocks, or rock are stronger. A benefit of using concrete blocks or rock for terraces is that excess rainwater can flow more easily through the gaps between individual stones to the next terrace level or to the street. Use shrubs, grasses, and perennials with deep root systems in terraces, which will help stabilize the soil and slow down water, allowing it to permeate the soil.

A perennial garden with a low terrace wall built of stone helps to prevent erosion and allows water to be absorbed by plant roots.

### *Permeable Hard Surfaces*

Using permeable surfaces for our driveways, walkways, and patios is an excellent option to consider around your home. They allow water to permeate through them instead of becoming runoff that needs to be collected elsewhere. A well-designed, waterwise landscape incorporates permeable hardscape surfaces throughout, adding interest and functionality while reducing water usage in the landscape. Materials such as brick, flagstone, permeable pavers, or gravel are options to consider. To allow water to permeate, brick, flagstone, or pavers have wider gaps filled with a porous filler such as small gravel that water can easily infiltrate instead of sand or concrete. A driveway entirely made of finely textured gravel (landscape rock) such as ¼-inch (6 mm) decomposed granite or pea gravel isn't a good choice since they compact over time, becoming more impermeable. However, using a plastic grid underneath the gravel prevents compaction and allows water to infiltrate through it. Gravel works well as a permeable material (without plastic grid support) on level surfaces for landscape areas that aren't compacted by vehicles driving on them.

A permeable driveway with pavers spaced farther apart and filled with fine gravel.

A popular type of channel in many gardens is dry creek beds (arroyos), which are shallow trenches lined with river rock or other decorative landscape rock. Their purpose is to move and slow runoff (typically from near a structure) through the landscape to allow water to soak into the soil. Their dimensions vary but average 3 feet (91 cm) in width and 12 to 18 inches (30–46 cm) in depth with gently sloped sides for a natural appearance. Unfortunately, many are just decorative or flow directly out to the street and don't harvest any water. A well-designed dry creek bed will curve through the landscape, slowing the runoff progression and allowing water to soak into the soil. An important note is the dry creek should have at least a slight slope to direct water away from the house. Locate plants alongside the sides of the dry creek bed where they can absorb water. Use river rock or

A dry creek bed curves through a desert landscape, moving water away from the house but slowing its progress to allow it to permeate the soil and water plants, reducing the amount that reaches the street as runoff.

other decorative rock that averages ¾ to 1½ inches (2–4 cm) in size to line the bottom of the dry creek, which prevents soil from washing out (erosion). Incorporate boulders along the edges for more decorative interest. These winding channels should end with a shallow basin where water can pool and soak into the soil instead of running into the street. The basin, or water catchment basin, should have a lower edge on the side facing the street to allow excess water to drain away once it's filled to prevent water from backing up.

### *Stormwater Planters*

In many urban areas, stormwater is routed from the street into planters alongside or in the middle of streets via an inlet cut into the curb. Water is channeled toward plants as it enters the planter. As discussed in chapter 4, plant roots help open the soil so water infiltrates more efficiently. This maximizes rainwater usage while reducing the amount of polluted water that reaches rivers and streams since the plants' roots and the soil help to clean the water by removing pollutants as the water travels downward into the soil.

Both active and passive water-harvesting systems can work beautifully together, and their use isn't limited to our homes and can be integrated around business properties, parking lots, and roadways. Harvesting rainwater results in healthier plants, a reduced water bill, cleaner water, and reduced risk of flooding.

## TURN A NONFUNCTIONAL DRY CREEK BED INTO A FUNCTIONAL ONE

If you have a dry creek bed that doesn't collect water, you can modify it into one that works. Ensure there is a source of rain runoff at the top of the dry creek, such as a downspout from a rain gutter or runoff from a porch, patio, or entryway. Make sure the dry creek bed slopes away from the house—regrade if necessary. Create a swale at the end of the dry creek bed to collect rainwater. The final step is to add plants alongside the winding bed and a shallow basin at the end.

Along a busy street in the desert Southwest, curb cuts in a median planter allow runoff from rain to flow through a channel that winds through plants, allowing the soil to absorb water and reduce the risk of flooding in the street.

Both active and passive water-harvesting methods are used within an Arizona back garden. A water tank collects water from the roof, and any overflow flows through underground pipes into swales surrounded by plants.

CHAPTER 6

# When and How to Water Plants Efficiently

**Access to water via rain, fog, or supplemental irrigation** is essential to growing plants—there are no plants that don't require moisture in some form to grow. Supplemental irrigation is commonplace if you live in a dry climate or if your garden has plants with higher water needs than rain will provide. In regions where rainfall is plentiful, the need to give extra water to plants may be a rare event unless you experience dry seasons or periods of drought. Using plants native to your region or from areas with similar rainfall patterns will reduce your reliance on providing supplemental water. However, most people's gardens are composed of a mixture of different plants with varying water needs.

A Colorado front garden filled with colorful drought-tolerant plants that receive supplemental irrigation during drier parts of the year.

Drip irrigation running down the sidewalk into the street due to a leak or from being turned on too long.

These two plants show evidence of overwatering due to being planted too close to a lawn. The bush morning glory (*Convolvulus cneorum*) has lank, straggly growth, while the tropical hibiscus (*Hibiscus rosa-sinensis*) shows signs of nitrogen deficiency.

## Common Watering Mistakes

Water is a precious resource, and once-reliable sources of water are shrinking, so we must use water wisely to ensure that we are watering efficiently. Before watering, we need to ascertain whether plants need water (on that specific day) to ensure we meet their water requirements, whether relying on rainfall, supplemental irrigation, or a combination of both.

As you walk through a garden, you are more likely to see watering problems than not. From too much or too little water, overly wet or bone-dry soils, irrigation leaks, sprinklers spraying the wrong area, or watering at the wrong time of day—there is no shortage of watering mistakes to discover in the landscape. Let's explore common watering mistakes and how you can avoid them.

### Overwatering

As a horticulturist and landscape consultant, I often encounter clients who are overwatering their plants as opposed to those who aren't watering enough. Overwatering is commonplace in most regions. Part of our human nature wants to care for our plants, and we often go overboard, giving them more than they need. In chapter 2, we discussed that water floods into the small spaces between soil particles, pushing out the oxygen when we water plants. Plants begin to suffer from a lack of oxygen if the soil stays wet too long. Additionally, watering too often flushes nitrogen from the soil, a vital nutrient plants need. Overwatering can lead to weaker plants that need frequent watering due to a shallow root system, resulting in an unhappy plant. If you tend to water plants too often, it's helpful to remember that plants have existed just fine on their own out in natural areas with no help from us.

How can you tell if you are overwatering your plants? First, look for signs of yellowing foliage. As nitrogen is flushed out of the root zone, the foliage will begin to show the effects of nitrogen deficiency with light green to yellow leaves.

Excessive watering can cause leggy growth and fewer blooms. In some instances, foliage will show signs of wilting, which is a sign that root rot is present that prevents roots from absorbing water. Perpetually moist soil that doesn't dry out or the presence of algae on the soil surface also can point to overwatering. Some plants are high-water-use and thrive in soil that is generally moist all the time, while others like periods of dry soil between watering. So, looking for symptoms such as yellowing and wilting can indicate plants are receiving too much water.

If you have an overwatered plant irrigated by rain only, consider moving it to a drier space in the garden, such as an elevated area or one that gets a bit more sun. In most cases, plants suffer from overwatering. The fix is more straightforward in this situation—don't water as often. For example, if you water a plant every three days, increase the number of days you don't water to five, six, or more. To determine the ideal frequency, stop watering and monitor your plant for signs of underwatering (see next section), and adjust your watering schedule accordingly. See page 93 to determine your ideal watering frequency.

## Underwatering

Problems with underwatering are less commonplace than overwatering, but it's essential to learn the signs to identify plants that need more water. Plants consist of 75 to 90 percent water; if they receive too little, leaves may wilt. Other signs include browning leaves that may curl. Plant growth may be stunted, with smaller leaves forming from a lack of water. Finally, the soil around plants will be dry and hard. Signs of underwatering typically show up more in warmer times of year.

We need to monitor our plants, particularly during summer or other dry times of the year, for signs of drought stress. Walk through your garden when it's being irrigated, and ensure water goes to plants. For plants that receive most of their water from rain, during times of less than average rainfall, check for signs of underwatering, and provide them an extra drink of water if needed. If your irrigated plants show signs of drought stress, make sure they are watered deeply each time, as watering too shallowly will cause the soil to dry out too quickly. You may need to increase the frequency of your watering schedule to ensure plants are getting the water they need.

### GOT WILTING PLANTS?

If you spot a plant that shows signs of wilted foliage, how can you tell whether it's caused by over or underwatering? The answer is simple: Look at the soil around the plant and lightly scratch the surface. If it's overly moist, the problem is too much water. Conversely, if the soil is quite dry, the plant is thirsty.

A drought-stressed natal plum (*Carissa macrocarpa*) with browning foliage and leaf loss due to insufficient water.

## Not Watering Deeply Enough

The key to a plant's vitality is found under the ground. Plants do best with deep root systems, and shallow watering limits how far down they can grow. Roots only grow in soil spaces that have access to moisture. Another problem plants face with shallow watering has to do with salts. Soils contain salts, and our tap water has natural salts dissolved in it. Dry climate regions tend to have higher amounts of salts in the soil. When plants aren't watered deeply enough, the water evaporates from the soil quickly, leaving behind the salts. Over time, this leads to a concentration of salts around the root zone, which isn't healthy for plants. You may see signs of browning leaf tips or a white powdery substance on the ground around plants, which indicates salt buildup in the soil. If your plants show signs of salty soil, water deeply for longer than usual to help flush salts down and away from the root zone of plants.

Other problems with shallow watering are roots more exposed to weather extremes, resulting in fluctuating soil temperatures. The soil dries out more quickly and must be irrigated more frequently. The root systems of plants are stunted when they aren't watered deeply, so they aren't as attractive or healthy as those watered to the appropriate depth. So, we must water plants to the correct depths to foster deep root growth (see page 90).

Salt buildup appears as a white powdery substance around this red yucca (*Hesperaloe parviflora*).

Using sprinklers in the middle of the day results in a large percentage of the water evaporating immediately into the atmosphere before it falls to the ground.

## Watering During the Wrong Time of Day

Drive down any street in the middle of the day, and you won't have to go far before you see sprinklers watering grass or other plants. So, what's wrong with that? Well, during midday or in the afternoon, 20 to 50 percent of the water a sprinkler shoots up evaporates instantly before it hits the ground. That is a lot of water wasted! Another problem with watering later in the day is that plants have more difficulty absorbing water if they have to deal with the stresses that a hot, sunny or windy day brings. This is true whether you use drip irrigation, a hose, or sprinkler. Additionally, using sprinklers in the evening or at night can cause problems by fostering the ideal conditions for fungal disease. So, that leaves just one ideal time for watering—early morning.

A therapeutic garden in southeastern Arizona with plants that have varying water needs. Three separate irrigation zones allow for the proper watering frequency for trees, shrubs, perennials, and succulents.

## One Watering Schedule Doesn't Fit All Types of Plants

Plants have different water needs. Their water requirements vary according to the type of plant: tree, shrub, perennial, groundcover, or succulent. To add to the variability, we add plants that produce fruit, vegetables, herbs, flowering annuals, and container plants. Then, there are higher water-use plants vs. those that are drought-tolerant. The problem arises when we irrigate all our plants with the same schedule. If you have a single irrigation line (valve) for all plants, by default, you need to schedule it to the requirements of your thirstiest plants. This results in the rest of your plants being either over or underwatered and, therefore, less healthy. Ultimately, this wastes water. The good news is that we can make modifications to ensure each type of plant is getting the right amount of water for their needs, which we will dive into later in the chapter.

## General Watering Recommendations

At this point, you may have realized watering plants isn't as simple as you think—especially if your goal is to use water as efficiently as possible. Variables such as the type of plant, the soil it's growing in, the method of watering, the water pressure, the depth, and the timing all play a part in ensuring plants get the water they need. If you feel overwhelmed, don't worry; we will break it all down to help you water your plants correctly.

### Benefits of Watering Deeply

First, let's discuss watering to promote deep root growth, which maximizes the amount of roots. The more roots a plant has, the healthier and more resilient it will be in dry conditions, which also has the benefit of a more attractive plant, which is what we want. The depth to which a plant's roots grow primarily depends on the plant type. Tree roots grow the deepest, while groundcovers and succulents have shallower root systems. Deep roots allow for more drought resilience and increased stability in windy conditions. Additionally, the further down roots grow, the more protected they are from temperature extremes, keeping them cooler in the summer months. Lastly, deeply rooted plants have access to more moisture in the soil and need water less often, which is especially vital in dry climate regions.

So, how do we encourage our plants to grow deep roots? It comes down to good soil texture, which we discussed in chapter 4, that allows water to permeate the soil, and how deeply we water. The path water takes through the ground opens it up for roots. In other words, roots will follow the flow of water. So, watering to appropriate depths is key to promoting deep root growth.

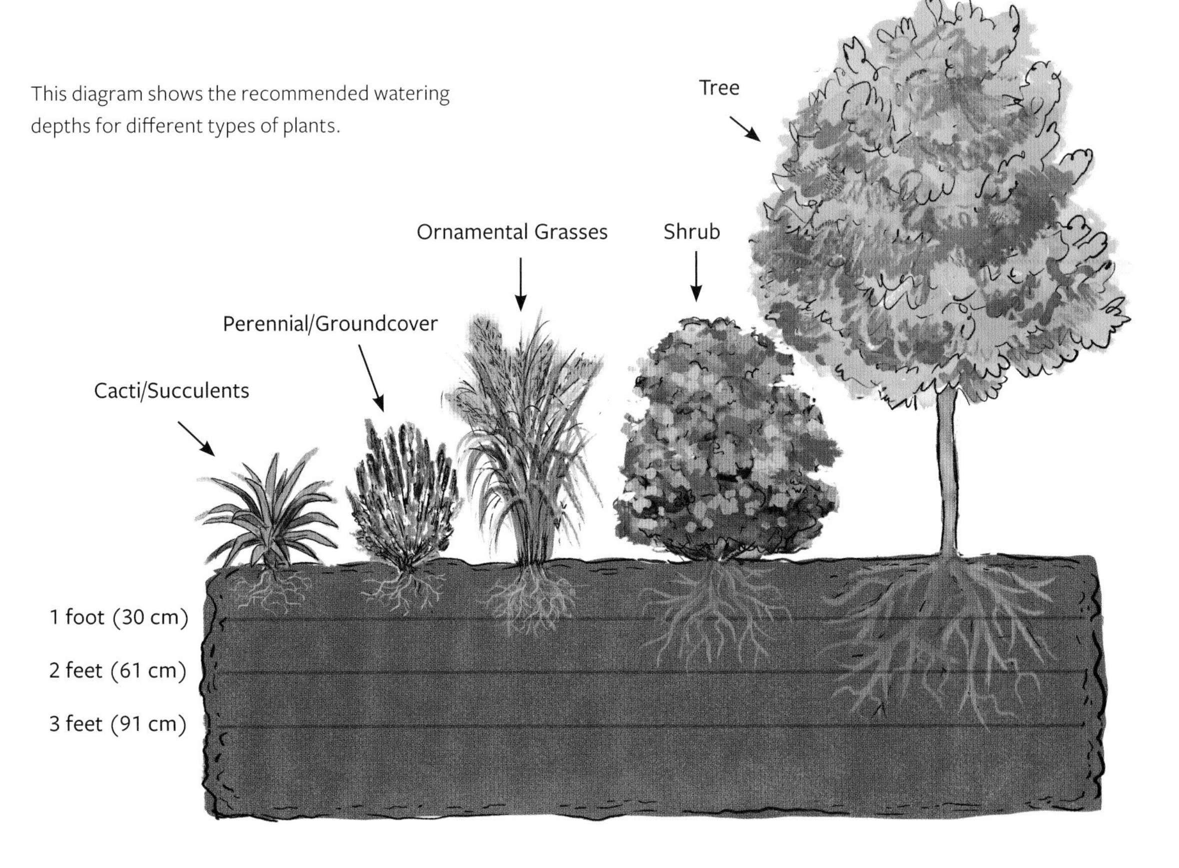


This diagram shows the recommended watering depths for different types of plants.

### *How to Determine Appropriate Watering Depths*

Now that we know how deeply we should water our plants, we need to figure out how deep we are watering each time. To do this, you will need a soil probe, a long screwdriver, or a length of rebar. An hour after watering your plants, push the soil probe into the wetted soil—it should go down fairly easily and will stop at the level where the downward progress of water has stopped. If you use a piece of rebar for the test, you may need to lightly hammer it into the soil. Pull the probe back out to see how deeply you have watered, telling you whether you need to extend your watering time and, perhaps, increase the water flow. If the depth is too shallow, you'll need to increase your irrigation time or apply more water each time.

Conversely, if you have watered plants too deeply, shorten the time water is on or reduce the water flow. If you have clay soils, it will take longer for water to permeate into the soil, while sandy soils will take less time. You don't need to perform this test on every plant in your garden but on two to three of each type of plant. Once you know the approximate time it takes to water to a plant's preferred depth, that amount of time won't change.

The author uses a soil probe to ascertain that her queen's wreath (*Antigonon leptopus*) vine is watered to the appropriate depth.

Various sizes of watering cans can make hand-watering easier, rather than having to carry large ones through the garden to plants.

## Watering Frequency

Once we have determined how long we need to water plants, it's time to determine how often we should water, over which there is a lot of confusion and inaccurate information. For this reason, usually plants receive too much water because people would rather err by watering too often rather than too little. To ensure that we are using water wisely, we need to figure out the right schedule for providing plants with the water they need, but no more.

Just as plant type and soil texture contribute to recommended watering depths, they also determine how often we should water our plants. However, other variables come into play with watering frequency, such as the climate a plant is growing in, whether a plant is a high-water-use plant or one that prefers drier conditions. Are plants being grown in full sun or shade, where their water needs may be less? What type of irrigation system are you using? Finally, the time of year comes into play—plants need to be watered more often in summer than spring or fall, while watering needs are at their lowest in winter.

In the author's garden, succulents like this 'Flying Saucer' (*Echinopsis* 'Flying Saucer') cactus have much lower water needs than the purple flowering groundcover behind it. 'Flying Saucer' does best with infrequent watering spring through fall.

A newly planted desert willow (*Chilopsis linearis*) tree, already connected to drip irrigation, receives extra water from the hose until it can be weaned to the watering schedule for existing plants.

***How Soil, Plant Types, Age, and Sun Exposure Affect Watering Frequency***

Let's explore these variables and how they affect your water schedule. First, sandy soils dry out more quickly than loam or clay soils, so water must be applied more often. On the other hand, clay soil retains moisture longer, so you can go longer periods in between watering. For plant types, the deeper a plant's roots, the longer it can go between watering. So, trees don't need water as often as smaller plants such as shrubs and perennials. Plants with shallow root systems, such as groundcovers, need frequent watering because their root systems are localized to the top foot of soil, which dries more quickly. The exceptions are cacti and other succulents, which can store water inside and don't need water as often. They prefer that the soil dry out completely for a while in between watering.

Delving deeper into plant types, we must determine whether a plant is drought-tolerant or requires more water to grow. Plants native to your region or those that come from regions with a similar climate likely won't need frequent, supplemental irrigation unless rainfall amounts are less than average. When we incorporate plants from areas where rainfall is more plentiful than ours, we will need to provide more frequent watering.

Newly planted plants have small root systems and need water more often than those in the ground for a year. The extra watering will help to promote root development and prevent the soil from drying out too quickly. The frequency of watering for new plants depends on the type of plant, the season, and the climate. Consult with your local nursery for recommendations for watering new plants. Monitor plants for signs of over or underwatering, such as yellowing foliage and dropping leaves, and modify the watering schedule as needed. After the first year, gradually increase the intervals between watering until you reach the recommended guidelines for your region for established plants.

Where your plant grows in your garden affects how often it needs water. If we take two identical plants and place one in full sun and the other in a shadier location, guess which one will need more water? Plants require more water when grown in the sun than in the shade, where they enjoy protection from the sun.

A Mexican daisy (*Erigeron karvinskianus*) needs less water in this Vancouver, British Columbia, garden in Canada than in drier regions such as Mediterranean climates, where it also grows.

## METEOROLOGICAL SEASONS AND WHY THEY MATTER

When changing how often we water our plants seasonally, we need to ditch traditional calendar seasons and rely on meteorological seasons instead. This alternate seasonal calendar begins on the first day of the initial month and is based on when seasonal temperature changes typically occur. Here are the meteorological seasons in the Northern Hemisphere: Spring: March, April, and May; Summer: June, July, and August; Fall: September, October, and November; and Winter: December, January, and February. We need to follow this calendar in regard to most gardening practices, especially watering.

### *Watering Needs Change Seasonally*

The final variable to discuss is how the weather affects a plant's water requirements. Not surprisingly, warm or hot summer temperatures are when plants need the most water. In winter, when temperatures are at their lowest, a plant's water requirements are minimal. For spring and fall seasons, water needs to land in between. One of the biggest watering mistakes is when plants are watered year-round on a summer watering schedule—even in winter. Adjust your irrigation schedule with seasonal weather changes. During a heatwave or periods of high winds, plants will appreciate more frequent irrigation. Put a reminder on your calendar when it's time to adjust your watering frequency to the usual timing so you will remember.

### *Where to Find Guidelines Specific for Your Area*

A valuable resource for how often to water your plants is likely at your fingertips. Many cities and towns have watering guidelines specific to where you live, created by experts, and largely reflect your region's specific climate and soil types. The guidelines are usually found on your city's website or through your water provider. Alternatively, a simple online search can pull up guidelines for your area. If you need help finding irrigation scheduling information for where you live, locate a larger city nearest to you with similar climatic and soil conditions for guidance or consult with your local nursery professional. Regional irrigation recommendations are an excellent starting point for setting up your irrigation calendar. It's important to note that these regional guidelines are general but a great starting point. You may need to adjust those guidelines due to specific conditions in your garden that differ.

### *How to Determine Your Ideal Watering Frequency on Your Own*

Another way to determine how often to water your plants can be done by observing your plants.

The primary way to achieve this is to observe your plants between watering cycles. Young plants have a smaller root system and will need irrigation more frequently than fully established ones that can go longer between watering cycles. A well-watered plant will have attractive leaves without signs of curling, browning, or wilting. Begin by watering your plants at the start of this test and then monitor them daily for signs of drought stress, such as wilting, browning foliage, or wrinkling in the case of succulents. Insert your finger (or shovel) into the soil a few inches to check moisture levels—if it is dry, plants likely need a nice drink of water. You know the plants need water once you see any of those signs. Ideally, we don't want our plants to get to this point every time, so shave off a few days from the non-watering days and set that schedule. Perform this test each season to determine the best irrigation intervals throughout the year for plants. For example, in the summer months, your shrubs show signs of wilting after seven days with no water. So, moving forward, you will ensure they are being watered every five days in summertime.

## SMART CONTROLLERS MAKE SCHEDULING WORRY-FREE

If the idea of fluctuating water schedules stresses you out or you wish you didn't have to remember to adjust your controller, there is another option. Smart irrigation controllers automatically adjust watering schedules in response to local weather—including temperature, wind, and rain events. They are typically installed by a landscape or irrigation professional, and they take care of the watering needs of your garden. Many controllers have wi-fi capabilities, making it easy to control them from your phone or computer.

### Irrigation Timers

Imagine a device that will turn your irrigation system on and off automatically, freeing you from having to remember to water or getting someone to water your plants when you are out of town. It is easy to forget to water your plants with the busyness of life and other tasks that demand our attention. You can take the hassle of turning the water off and on with an irrigation timer, also called an irrigation controller. Plants watered on a schedule tend to be healthier than those watered inconsistently, which can stress them. There are different kinds of irrigation timers, beginning with a simple battery-operated timer that connects to your outdoor faucet and will control one watering line. More advanced irrigation timers are connected to electricity and mounted on a wall. They connect to multiple irrigation lines (zones) and control the day, the time water is applied, and for how long. These timers are connected to your home's electric system and are useful for gardens with multiple irrigation lines. Installation of an irrigation timer is typically done by a landscape professional who connects the timer to your electric system and to underground irrigation valves.

### Why Separate Irrigation Zones Are Best

Earlier, we discussed that plants had different watering needs, some needing more than others.

Many irrigation systems around a home consist of a single irrigation line (valve) that waters different kinds of plants on the same schedule. A one-size-fits-all approach to irrigation is where we can run into problems with overwatering because, by default, we need to water on the schedule of the highest water-use plant. A water-smart irrigation system consists of several individual watering lines—each serving plants that share the same water requirements regarding scheduling. For example, a home has three individual irrigation lines providing precise water scheduling for trees, shrubs/vines/groundcovers, and a vegetable garden. The more we can tailor our irrigation systems for different plant needs, the more you will enjoy healthy plants, reduce water waste, and have a less expensive water bill.

However, what can you do if separate water lines aren't feasible? If you have a drip irrigation system, add individual drip emitters that you can turn off and on for certain plants as needed. Additionally, you can provide extra water to your thirstiest plants with a hose. Finally, you can use portable drip irrigation vessels for plants that you can make yourself (see page 162). If your irrigation system consists of you watering by hand, concentrate your thirstiest plants near the house where they are easier for you to water.

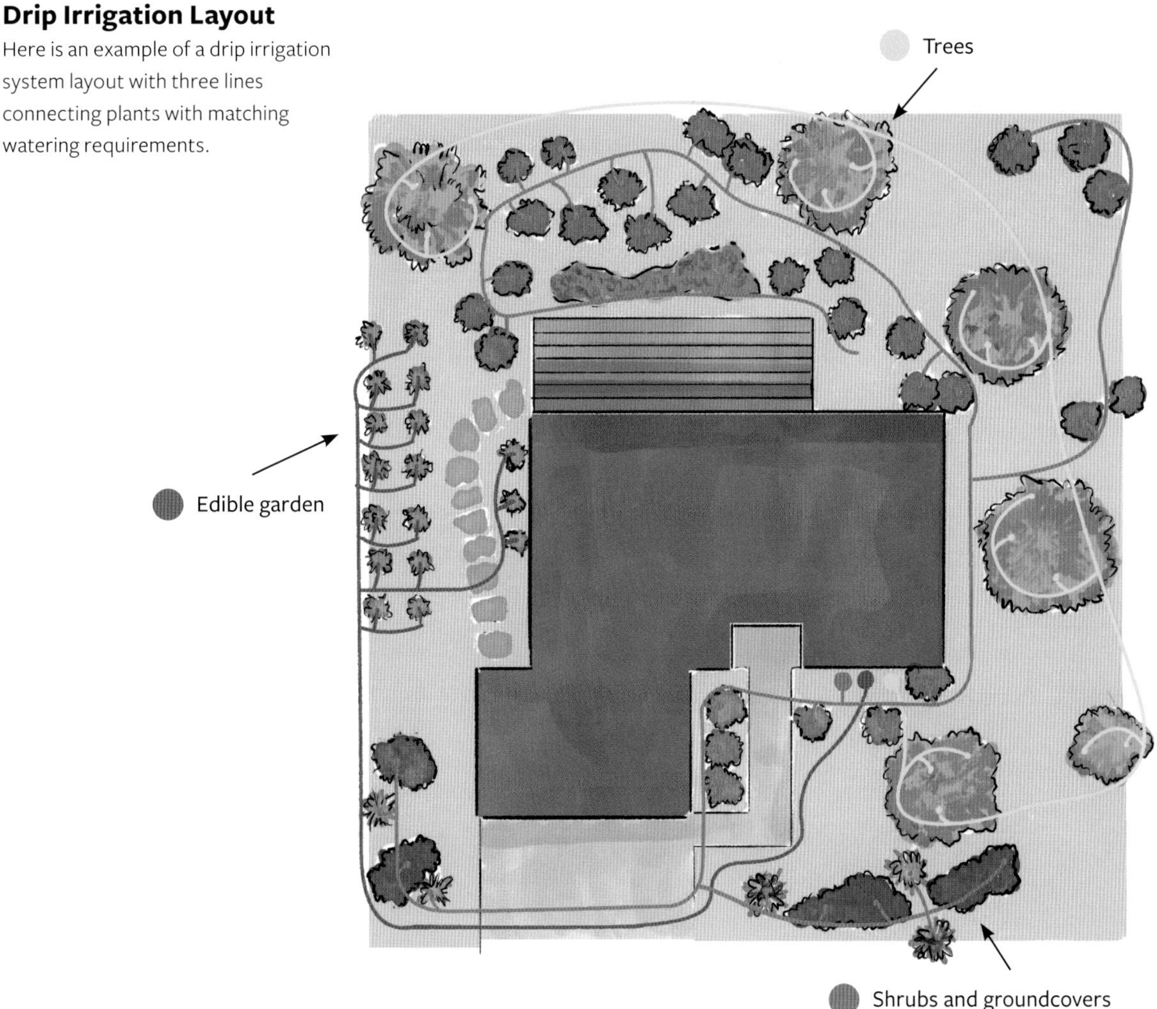


**Drip Irrigation Layout**

Here is an example of a drip irrigation system layout with three lines connecting plants with matching watering requirements.

## Applying Water at the Root Zone

Where you apply water around plants matters, and many people get it wrong. We need to provide water to the root zone of plants—and there is often a misunderstanding about where the root zone is, which gets people into trouble. When we add a new plant, we water it right next to the base of the plant, which works because that is where the roots are. However, as a plant grows, so do its roots, expanding outward, often past the edges of the plant itself. Most of a plant's water-absorbing roots reach out to the plant's dripline, which is the plant's outer edge, where water needs to be applied. As a plant grows, the location of the dripline changes, moving outward until the plant reaches its mature width. So, where we apply water must move too. If you use drip irrigation to water plants, drip emitters must move with a plant's growth. For plants with basins around them for watering, the basins need to be enlarged, extending to the plant's dripline.

Another factor to consider is that as a plant grows larger, its water needs become greater as it requires more water to survive. I didn't say they require more frequent watering, just more water applied each time to the depth appropriate for that type of plant. In general, plants need less frequent watering the older they become.

# Efficient Watering Methods

There are different techniques used to water plants, from the humble garden hose to sprinklers, all the way to a drip irrigation system, and there are pros and cons for each method, each with varying levels of water actually reaching plant roots. Let's explore popular ways to water plants, beginning with the most efficient methods.

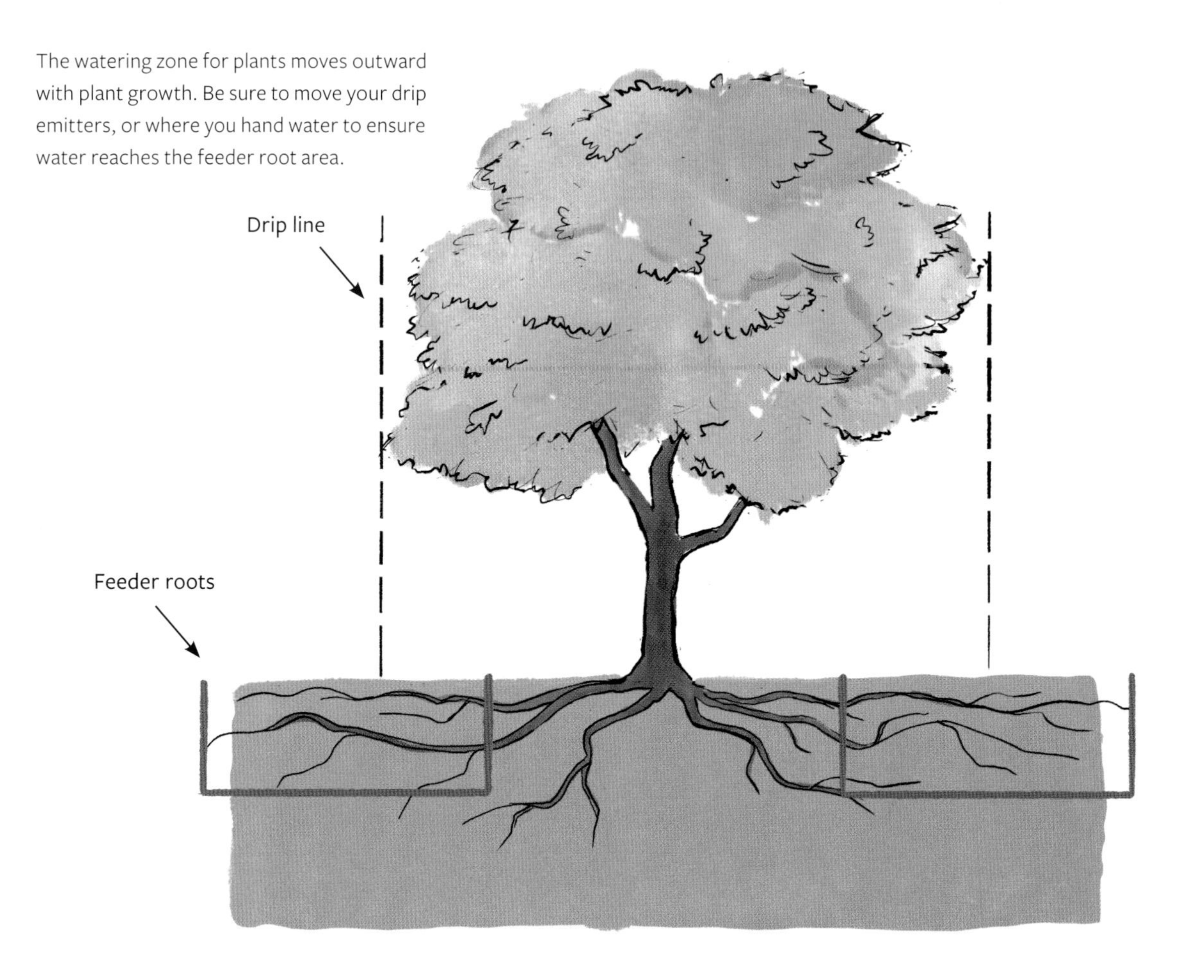


The watering zone for plants moves outward with plant growth. Be sure to move your drip emitters, or where you hand water to ensure water reaches the feeder root area.

## Drip Irrigation

A relative newcomer to the gardening world, modern drip irrigation began its journey in the deserts of Israel in the 1950s when an engineer and inventor discovered a tree growing better than the surrounding plants. With no known water source, the discovery of a small leak in an irrigation pipe near the tree inspired the invention of modern drip irrigation. Since then, drip irrigation has rapidly gained in popularity, particularly in dry climate regions where water supplies are limited. The increasing prevalence of drip irrigation is due to its high efficiency, with up to 90 percent of water released going directly to the plant. Water is slowly delivered to plants by drips, preventing runoff and allowing water to permeate the soil. Drip irrigation is the most efficient of all irrigation methods.

You can connect a drip system to your outdoor hose faucet using a battery-powered timer.

### *How Drip Irrigation Works*

Water runs through plastic tubing interspersed with emitters that "drip" water out directly to the base of plants, where it infiltrates into the soil to the root zone. This dripping action makes this type of irrigation so efficient since there is little water lost to evaporation, and most of the water reaches the roots of plants. Water goes to the specific area where roots reside, and the bare areas between plants remain dry, which saves water and can help prevent weed growth. If you garden on a slope, the ability of drip irrigation to slowly deliver water to the root zone makes it a preferred method for watering plants without runoff. Drip irrigation also helps to prevent fungal diseases since irrigation water doesn't wet the foliage of the plants. This waterwise irrigation technique is a popular option for watering trees and other plants throughout the landscape. It is also an excellent option for edible gardens and container plants.

The water source for drip irrigation can be an outdoor water faucet or in-ground irrigation valve. DIY drip irrigation kits are available with all the components for installing drip irrigation yourself that attach to an outdoor faucet. A landscape

A drip irrigation emitter slowly drips water to a shrub in the author's garden.

## HOW DRIP IRRIGATION WORKS

*Drip irrigation systems consist of:*

**Poly Tubing:** Also referred to as mainline tubing, this flexible, plastic tubing is ½ to ¾ inch (13–19 mm) in diameter and is placed around plants throughout the landscape. It is buried or lies on the surface of the ground and connects to a faucet or in-ground valve (water source). Emitters, or small holes, are interspersed along the length of tubing that water drips out of. PVC is an alternative to poly tubing (see sidebar on facing page).

**Backflow Preventer:** This component keeps water from the irrigation system from backing up into the household water supply line.

**Pressure Regulator:** A pressure regulator lowers the pressure of incoming water into the drip irrigation system, which works under low-pressure conditions only.

**Filter:** A filter prevents debris from entering drip tubing, which can clog up the system.

**Connectors:** Connectors attach sections of mainline tubing together. Barb connectors connect ¼-inch (6 mm) drip tubing to mainline tubing.

**Drip Tubing:** Also referred to as "microtubing" or "spaghetti," drip tubing is thin tubing that is ¼ inch (6 mm) in diameter that attaches to the poly tubing with an emitter on the other end.

**Emitters:** Emitters are small plastic devices that release small amounts of water to the base of plants. Emitters are often color-coded based on how much water they release, ranging from ½ gallon per hour to 6 gph (0.03–0.4 Lpm). Adjustable emitters allow for adjusting the water flow or shutting off an individual emitter. Emitters are either connected directly to the poly tubing or to the end of the drip tubing that runs from the mainline tubing.

**End Caps:** At the end of mainline tubing, an end cap is fitted to prevent water from flowing out of the end of the poly tubing.

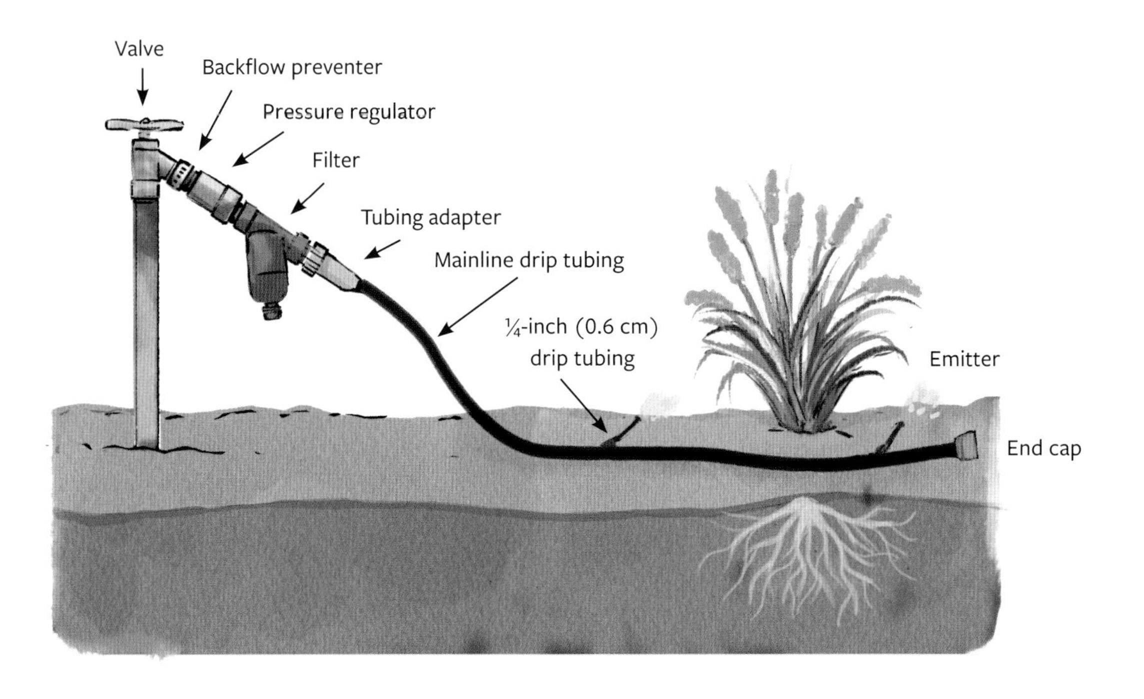

## PVC VS. POLY TUBING DRIP IRRIGATION: WHICH IS BETTER?

When installing a drip irrigation system, you may face two choices of materials for the system. Systems with poly tubing (or drip tape) are most popular in a garden setting as they are flexible and generally less expensive. Drip irrigation kits that come with all the components of a drip irrigation system utilize poly tubing. PVC is made from vinyl irrigation tubing, which, unlike poly tubing, is rigid. It is stronger than poly tubing, less susceptible to developing leaks, and has a longer life span. PVC is a better choice if you have animals that like to chew into your irrigation system. However, it isn't flexible, so more fittings are needed to configure it around plants. Additionally, PVC is more susceptible to damage from freezing and thawing, so a poly tubing system is often preferable in cold winter locations. Consult with an irrigation professional in your area to see which option best suits your needs.

professional often does the installation for a more complex system with multiple lines and in-ground irrigation valves.

### *Drip Emitters—Which Size Should You Use?*

Drip emitters come in different sizes that release varying amounts of water, which allows you to customize the amount of water released for different types and sizes of plants. In general, the larger the plant and the deeper its roots grow, the larger-sized emitter it will require. Drip emitter sizes range from ½ gph (gallons per hour) (0.03 Liter per minute), 1 gph (0.06 Lpm), 2 gph (0.13 Lpm), and 4 gph (0.25 Lpm). In the table below are general guidelines for drip emitter size and number. However, higher water-use plants may need more.

Adjustable emitters are another option, with a small dial on the top, allowing you to adjust the amount of water that flows out or simply turn off. They are a useful option for drip irrigation systems where plants with varying water needs are on the same line.

### *Drip Irrigation Maintenance*

While drip irrigation is the superior choice for waterwise gardening, it does have a drawback due to the small holes that water travels through. Minerals from water can build up and plug holes in drip emitters after a while. However, the solution is easily solved by replacing the old emitter with a new one. Flushing of the drip irrigation system two to three times a year can minimize mineral buildup. To do this, remove the end cap from the poly tubing and turn on the water for two minutes, which will help remove accumulated minerals. Another maintenance task is to clean out the filter near the water source of any debris. Use water from the hose or a brush to remove debris from the filter. While your drip irrigation system is running , twice a year walk around inspecting each emitter to ensure it is working correctly to water plants. You will either see water dripping from the emitters or wet spots by each plant. Individual emitters can become clogged without you noticing until you have a dead plant, so doing this visual inspection is essential,

A clogged drip emitter due to mineral buildup from hard water, leading to the decline of the plant it was watering.

| PLANT TYPE | DIAMETER (FEET) | NUMBER OF EMITTERS | FLOW RATE |
|---|---|---|---|
| Cacti/Succulents | 1–3 (30–91 cm) | 1 | ½ gph (0.03 Lpm) |
| Groundcovers/Perennials | 1–3 (30–91 cm) | 1 | 1 gph (0.06 Lpm) |
| Shrubs | 3–6+ (91–1.8 m) | 2 | 2 gph (0.13 Lpm) |
| Small Trees | 6–10 (1.8–3 m) | 2 | 4 gph (0.25 Lpm) |
| Trees | 11–14 (3.4–4.3 m) | 3–4 | 4 gph (0.25 Lpm) |
| | 15–20 (4.6–6.1 m) | 3–4 | 4 gph (0.25 Lpm) |
| | 21+ (6.4 m) | 5–10 | 4 gph (0.25 Lpm) |

## STILL WATERING A NONEXISTENT PLANT?

We often forget to plug irrigation when we take out a plant. To avoid wasting water that is no longer needed, cut off the drip emitter and plug it using a "goof plug" when you remove a plant. If you add a new plant in the future, you can cut off the goof plug and insert a new drip emitter.

and you can replace a blocked emitter. I highly recommend keeping drip emitters aboveground (instead of attached directly to the poly tubing) when possible, which makes them easy to replace and spot any problems with mineral buildup.

### Soaker Hose

A soaker hose is a porous rubber hose where water seeps through thousands of tiny pores along its entire length. Similar to drip irrigation, soaker hoses slowly drip out water to the base of plants. They are an excellent option for providing water in certain areas around the garden as they are portable and available in varying lengths—from 25 to 100 feet (7.6 × 30 m). Connect the soaker hose to an outdoor faucet (spigot). Soaker hoses need very little pressure, so you don't have to turn the water on full blast. They are a great option as they cost less than a drip irrigation system and can connect to a hose-end timer. Their flexible length allows you to weave them around plants. Cover with a layer of mulch to protect from the sun and preserve moisture. You can move a single soaker hose from zone to zone in the garden or purchase several and keep them in place, bringing the regular garden hose out to connect to each one. To test how long to water with this method, turn it on for 30 minutes and then use a soil probe to see how deeply the water has penetrated and adjust the time as needed.

A soaker hose slowly drips water to purple trailing lantana (*Lantana montevidensis*).

While soaker hoses are efficient, they waste more water than drip irrigation because water delivery is focused on more than individual plants since water is released along the entire hose length.

Unlike drip irrigation systems, you cannot adjust how much water is released for specific plants with a soaker hose. They are best suited for level areas as they provide uneven watering on sloped areas. Over time, leaks appear along the hose that spurts water out. Mineral buildup can block the pores, and direct sun will break down the rubber material of the soaker hose. The lifespan of a soaker hose is typically two years.

## Ollas

This practice of irrigation consists of burying earthenware containers made of unglazed clay or terracotta alongside plants. The vessels are filled with water that slowly leaches into the surrounding soil, providing water for plant roots. Commonly referred to as *ollas* nowadays, the use of buried earthenware vessels to water plants was practiced thousands of years ago by ancient farmers in China, the Middle East, and in the Americas and is still used today. Plants respond to even soil moisture by growing a healthy root system and don't have to endure the stress of dry periods. Another benefit is that since plants are watered from underneath the ground, weed growth is decreased.

Most of an olla is buried in the ground, with the top couple inches remaining aboveground. They come in different sizes, depending on the water requirements and root depth of the plants they are used for. Ollas have a hole at the top with a removable lid to prevent mosquitos and water from evaporating. As moisture slowly seeps into the soil, plant roots grow toward the olla as they pull water from it. Because the plant's water source sits underground, little water is lost to evaporation. Water takes a few days to a week to empty from the olla; depending on the season, water demands of the plants, and soil type, smaller ollas will empty more quickly and need to be replenished with water more often than larger ones.

Ollas have resurged in recent years as a time-tested irrigation method.

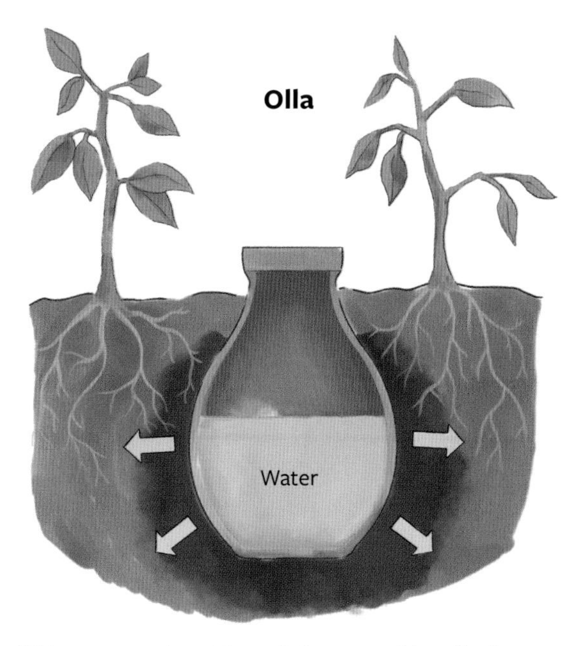

Water moves outward through the permeable walls of an olla as plant roots absorb moisture from the soil.

### *How to Use an Olla*

Ollas are a popular option for edible crops and container plants but are also suitable for other types of plants, except for succulents. If you have heavy clay soil, amend the soil with compost to improve drainage before adding an olla (see chapter 4, page 61). To install, dig a hole that can fit the olla's widest point, allowing the top 2 inches (5 cm) of the olla to remain above the soil level. Locate plants within 2 to 3 feet (61–91 cm) of the olla for adequate watering—closer for smaller plants. Avoid placing an olla near tree roots or large shrubs whose roots may crack the olla. Dry soils will deplete the olla reservoir quickly, so refill more often for the first three to four weeks until the soil becomes evenly moist and roots migrate toward the olla. Larger ollas can be spaced further apart than small ones. In general, space ollas every 3 feet (91 cm) if using large ones. For young plants with shallow root systems or newly sown seeds, you must provide supplemental irrigation until roots reach deeper to access moisture from the olla. You can purchase ollas, but you can make your own inexpensively (see how in chapter 11, page 164).

Check the water level in your ollas every other day to see if it needs to be refilled. In the summer months, you may need to add water every other day, while in spring or fall, once to twice a week may be sufficient.

### *Caring for Ollas*

For ollas made from unglazed terracotta, some maintenance is required to keep them functioning optimally. Salts from water and soil can build up on the surface of ollas, which create a white residue. Periodically soak in distilled water and lightly brush off the salty residue as needed. While some people add water-soluble fertilizers to their olla, this can contribute to residue buildup on the surface and hinder water movement into the soil. It's best to apply fertilizer separately and not add it to the olla. Ollas can develop cracks in freezing weather. If you live in a region with cold winters where freeze and thaw cycles are normal, dig up your ollas to protect them through the winter and rebury them in the spring.

## Watering by Hand

Using a hose or watering can has been the most common method for watering plants for ages. However, it doesn't rank high for water efficiency. Additionally, using a hose involves wrestling an unwieldy hose, pulling it through the garden, getting it stuck on plants or rocks, and developing kinks that stop water flow, only to reel it back in once you have finished watering. If a watering can is your preferred method for watering, there is the labor involved of filling it up and carrying it through the garden to its destination and back again to fill up. However, many people find hand-watering calming, and it can be a useful strategy for watering containers, new plants, and periodic deep irrigation of certain plants.

Issues arise when irrigating with a hose because we don't know exactly how much water a plant receives—is it too much or too little? Runoff is

A rain wand is a gentler way to water plants from a hose, without disturbing soil and lessening runoff. Be sure to use one with a shutoff valve to prevent water loss as you move from plant to plant.

Sprinklers not configured correctly result in overspray onto the sidewalk.

another problem since water comes out too quickly to be absorbed by the soil, so it flows away from the plant without permeating down to its roots, resulting in shallow-rooted plants. Water is wasted in the spaces between plants as you move from one plant to the other. Additionally, if the only water your plants receive comes from a hose instead of an irrigation system with a timer, they depend on you to remember to water them. Unexpected events, busyness, or vacations may cause your plants to suffer from underwatering.

If a hose is your primary irrigation tool, focus on watering where the plant's root zone is, which helps to reduce water loss from evaporation. We want to avoid wetting the leaves of plants, which can lead to fungal diseases, so be sure to water from the base of the plant. A hose can be a helpful method for plants that may need periodic deep watering when set at a slow trickle of water at the root zones of plants and allowed to water for 30 to 60 minutes.

## Sprinklers

One of the most common methods people use to water is also the most inefficient. Sprinklers are how we water lawns and, in some cases, other plants. Water from the sprinkler is sprayed up into the air and falls back to the ground in tiny droplets, mimicking the action of rainfall. Water is released evenly over plants, eventually making its way to the root zone. In-ground sprinkler systems consist of underground pipes and sprinkler heads. Sprinkler heads are configured to release water over a specified area, from a full circle, half a circle, to a smaller width. Other types of sprinkler irrigation are aboveground, with water released from a sprinkler attached to the garden hose. These are portable and can be moved to different areas. This method is more labor intensive.

The upward action of spraying water into the air before it hits the ground means that approximately 20 to 50 percent of water from sprinklers evaporates immediately, or shortly afterward, as it sits on top of plant leaves. In dry climate regions, this percentage can be even higher. Another disadvantage of sprinkler irrigation is that fungal diseases are more prevalent with sprinkler usage due to wet foliage. While using sprinklers is the preferred method to water lawns, it isn't for other plants, such as trees, shrubs, vines, and groundcovers. If sprinkler irrigation is how you water plants in your landscape, consider transitioning to drip irrigation, which is much more water efficient.

### *Best Watering Practices for Lawns*

Lawns are thirsty, and sprinkler irrigation isn't very water efficient. If you have a lawn, the biggest water impact you can make is taking it out. You can begin by removing nonfunctional lawns (decorative only) or reducing the size of your lawn, which will go a long way to reducing your water usage. However, what if lawn removal isn't an option for you? Some neighborhood's require a green lawn, or it could be that others in your household aren't willing to get rid of the lawn. So the sidebar on the next page explores how you can water your lawn to ensure you water as efficiently as possible to decrease water waste.

The removal of a large section of a front lawn creates curb appeal with plants that use less water, including agave (*Agave* spp.), prickly pear cactus (*Opuntia* spp.), yellow emu bush (*Eremophila glabra*), and angelita daisy (*Tetraneuris acaulis*).

## WATER-EFFICIENT PRACTICES FOR LAWNS

- » Ensure that sprinkler coverage is accurate and falls on the grass with no overspray onto other surfaces. Adjust sprinkler heads as needed.
- » Apply water so it has time to permeate the soil without runoff. You may have to split the irrigation time into shorter segments to allow water to soak into the soil so it doesn't runoff.
- » Water early in the morning, which decreases the amount of water that evaporates quickly.
- » Avoid watering too often, which leads to fungal disease and soil compaction. Recommended guidelines are to water lawns twice a week when they are actively growing—in cooler or rainy times of the year, grass needs less frequent watering. Consult regional watering guidelines for more specific instructions.
- » Water to a depth of 6 to 8 inches (15–20 cm) each time to promote deep rooting. You can test the depth with a soil probe to determine how long to turn water on.
- » Another option for knowing when to water your lawn is to look for signs of wilting and then adjust your watering schedule to water two days before wilting typically occurs. This timing will change with the weather, with grass needing more frequent irrigation in summer.
- » Skip watering when you receive ½ inch (1 cm) or more of rain. Consider installing a smart irrigation controller, which can detect when rain falls and will automatically skip an irrigation cycle.
- » Check your irrigation system for broken sprinkler heads, leaky valves, or underground leaks annually.
- » Allow your grass lawn to grow taller, to 3 inches (8 cm), providing more shading to the ground and preventing it from drying out too quickly. When you mow your lawn, don't remove more than one-third of its total height to reduce stressing the grass and increasing its water needs.
- » Skip overseeding for winter lawns since dormant lawns use very little water. In milder winter climates, lawns are often overseeded to provide a year-round green lawn, which requires a lot of water, particularly when the seed is growing in. It's actually healthier not to overseed your lawn, and it saves water.

Irrigation pipes can crack underground, leaking water. Look for wet spots or a higher water bill, indicating a leak in your irrigation system.

An adjustable drip emitter that has popped off its head.

## Check Irrigation Systems Periodically for Leaks and Other Problems

While monitoring your plants for signs of stress or disease is important, it's also crucial to keep an eye out for potential problems with your watering system. As the components of your irrigation system age, such as your valves, irrigation lines, sprinkler heads, and drip emitters, they can develop problems.

### *Watering on a Slope*

A problem you may run into is irrigation water running down the sidewalk or a slope before the soil can absorb it. Or, if you have basins around trees to allow for deep watering, water may spill out before reaching the appropriate watering depth. Thankfully, the solution is easy—split the amount of time water is turned on into two or three shorter cycles, allowing water to absorb without runoff.

### *Leaks*

Irrigation systems can develop leaks at the valves that connect the system to the water source or along irrigation hoses or PVC pipes. Signs of leaking include water inside the valve box, indicating a leaking irrigation valve. Walk through your garden at least twice a year, looking for unexplained wet spots, which point to a crack in irrigation piping leaking water or a sprinkler head that has popped off. Sometimes, there are no apparent signs in the garden of a leak, but an extra high water bill may indicate a hidden leak underground.

### *Drip Emitter Problems*

Drip irrigation emitters can sometimes pop off, causing water to stream out. Two to three times a year, go out into your garden while the drip irrigation is running and make sure no emitters have come off and are spouting water. It's a good idea always to have extra drip emitters ready to replace any that have popped off in the garden.

### *Winter Prep for Irrigation Systems*

In cold winter regions, all types of irrigation systems must be blown out to remove all water to prevent any splits or breakage of irrigation system components that occur when water freezes. Before the arrival of freezing temperatures, any water left in irrigation pipes, tubing, valves, and all other parts must be removed. A landscape professional can do this for you, or you can learn how to do it yourself with available online tutorials.

CHAPTER 7

# Maintaining Low-Water Plants the "Right" Way

**An attractive landscape** around your home requires maintenance to look its best. When you ask homeowners what garden tasks they perform, the usual answer is mulching, pruning, watering plants, and perhaps fertilizing. The level of maintenance people dedicate to their gardens depends on several factors, including the types of plants they have, whether they are well-suited for the climate, and individual preferences. You can have a garden that requires frequent maintenance if you have plants that need extra fertilizer and water to survive the climatic conditions present in your garden and nutrient levels in the soil. Gardeners who prefer a neat and tidy garden with neatly trimmed plants and no leaf litter or fallen flowers on the ground will have to dedicate themselves to frequent maintenance.

Decomposed granite acts as an inorganic mulch covering this dry climate landscape that allows water to permeate into the soil underneath.

Conversely, you can enjoy a beautiful outdoor space that requires much less maintenance by selecting plants that thrive in your climate with minimal to no supplemental watering. The exception is if you live in a dry climate region where regular applications of water may be required unless you grow only native plants. Choosing the right plants for your soil type will mean that amendments are seldom needed to modify the soil. Additionally, selecting plants that thrive with the nutrient levels already present in your soil means little to no fertilizer is required. Finally, if we allow plants room to grow to their natural size and shape, we experience the texture and stature they bring to our gardens with infrequent, strategic pruning practices.

If you had to guess which type of garden is more waterwise, you'd be correct in choosing the one that requires less maintenance. In most instances, the more we maintain our plants, the more we increase their water needs. Let's explore our maintenance tasks to ensure that we use methods that conserve water.

## Mulch, Mulch, Mulch

Some maintenance practices have a more significant impact on decreasing our water usage, with mulch residing at the top. Mulch is a soil covering around plants that provides several benefits. First, mulch prevents soil from drying out too quickly by providing protection from the sun and wind. Mulch allows plants time to absorb water and increases the interval between watering, either by rain or supplemental water. Another benefit of mulch is that it reduces weeds while insulating the soil from cold and hot temperatures, which leads to happier plant roots.

There are two types of mulch, organic and inorganic—both provide welcome benefits for plants, but there are marked differences between the two. Let's explore those differences so you can select the type best for your garden and water savings.

### Organic Mulch

From a water conservation standpoint, organic mulch is the best choice for decreasing water usage for plants as it improves soil structure, which is vital to maximizing the water efficiency of plants. Organic mulches are the most commonly used mulches and are made of materials such as bark, leaves, natural wood chips, pine needles, straw, and compost. The benefits to plants from this type of mulch include improving the soil's ability to hold onto the right amount of water. Over time, as organic mulch breaks down, it supplies low levels of beneficial nutrients for the soil while also promoting deeper root growth. This mimics what happens in nature as plants shed leaves, which act as a mulch and add nutrients to the soil when they decompose. Another benefit is that organic mulch comes from renewable resources, such as ground wood chips from your local arborist, which they often offer for free.

To maximize the water efficiency of plants, apply organic mulch in a 2- to 3-inch (5–8 cm) layer around the root zone of plants. Keep mulch from touching the base of trees or shrubs by keeping it about 6 inches (15 cm) away to prevent fungal disease from developing. As organic mulch breaks down through the year, improving the soil, reapplication is required in spring or fall.

### Inorganic Mulch

Landscape rock (gravel) is an inorganic mulch that is very popular for landscapes—especially in dry climate regions. Rock mulch comes in different sizes and colors, and in addition to the water-conserving benefits it provides, it is used as a decorative element in the garden. Inorganic mulch prevents the soil from drying out too quickly and helps moderate soil temperatures to a lesser degree than organic mulches. Landscape rock lasts longer than organic mulch and doesn't need to be replaced annually. Over time, the rock will appear to thin

## ARE YOU THROWING AWAY FREE MULCH?

Instead of buying bags of mulch, you may have all you need to mulch your garden. In nature, plants make their own mulch from the leaves, bark, and flowers they drop. We often see that as plant litter to clean up with a blower or rake. Instead, allow plants to mulch themselves by allowing leaves to remain or by directing fallen tree leaves around them. When using leaves as mulch, they must be shredded into smaller pieces, which can be done with a mulching lawn mower, leaf blower vacuum, or a shredder.

Pink ice plant (*Delosperma cooperi*) surrounded by rock mulch in a Colorado garden.

A garden in Perth, Australia, with wood chips surrounding a native pink rice flower (*Pimelea ferruginea*).

Flowers and plant leaves help to moderate the temperature of landscape rock in summer while filtering down into the soil as they break down, improving the soil and decreasing green waste.

out, exposing bare soil due to soil building up over the rock. Typically, you will need to apply a new layer of rock every five to seven years.

Apply landscape rock at a 2-inch (5 cm) layer. Because rock is a nonliving material, you can place it up against plants. Using landscape rock as a mulch won't improve the soil because it doesn't break down like organic mulch or add nutrients. Gardens with a rock mulch tend to be warmer because rock absorbs and re-radiates heat from the sun, which can increase plant water needs—especially for those with higher water requirements.

If you live in a region where landscape rock is prevalent or required in some neighborhoods, you can enjoy some of the benefits of organic mulch. Allow fallen plant litter (leaves, flowers) from shrubs and groundcovers to remain underneath the plants instead of blowing or raking them away. Allowing even a shallow layer of organic matter over the rock disrupts its heating effect, and plant debris can make its way slowly down through the rock as it breaks down to help improve the soil.

Rubber mulch is an inorganic mulch you shouldn't use in the garden. This type of mulch is made from recycled tires and comes in different colors that can resemble organic mulches. The rubber decomposes slowly and only needs replacing every ten years, which is why it is popular. However, rubber mulch releases toxic amounts of heavy metals and hydrocarbons as it breaks down, which raises health concerns, so it's best to avoid using it in the garden.

## Water-Efficient Pruning Practices

How we prune our plants has a direct effect on their water needs. The more we maintain our plants, the more we increase the water they require. Homeowners typically have one of two preferences regarding how their plants look. One is the garden with plants shaped into balls, squares, rectangles, or sculptured shapes. This formal style gained popularity in the 1800s during the Victorian era and is still preferred by many today. The other style is a landscape with plants allowed to grow into more natural shapes. As you can guess, the first type of garden takes much more pruning to keep plants in geometric shapes. But did you know that the more often we prune plants, the more their water needs increase? Let us explore the reasons why pruning makes plants use more water.

### Why Frequent Pruning Increases Water Use

The beautiful shades of green foliage we appreciate in our garden are how plants make food for themselves. Leaves turn the energy from the sun into food and store water and energy for plants. When a plant is pruned, it loses foliage that it has to regrow to replace what was lost, which requires more water than usual. For plants frequently pruned to maintain formal shapes or to keep them from outgrowing their too-small space, this creates stress for the plant as it devotes most of its resources to constantly regrowing leaves only to have them pruned away again shortly after. If you frequently prune your plants, their lifespan may be shortened due to the cycle of constantly having to regrow leaves to replace those lost. Additionally, you lose the natural shape and texture that can beautify your garden.

Cape honeysuckle (*Tecoma capensis*) is allowed to grow to its natural shape with annual pruning in spring.

Formally pruned shrubs that rarely flower with signs of dieback due to excessive pruning.

The author prunes a wayward branch from a pink trumpet vine (*Podranea ricasoliana*) that is growing in its natural shape.

### Why Less Pruning Is Best

Periodic pruning benefits plants' health by removing diseased portions, weak growth, and deadheading spent blooms of perennials. Pruning once or twice a year (or less) can be all that is needed to promote healthy plant structure, which can consist of pruning wayward branches and removing dead growth or old, unproductive branches of shrubs. Severe pruning can be done for certain plants, shrubs, or perennials to rejuvenate and stimulate new growth. In my garden, I have flowering shrubs that I prune back severely every three years and groundcovers and trees that are pruned annually. The pruning needs of individual plants vary, and you should reach out to a local nursery professional or master gardener in your region for guidelines. If your garden maintenance is performed by a gardener (landscaper) whose default is pruning monthly during the growing season, take them off autopilot and instruct them on how and when you want your plants pruned.

Plants are okay with being pruned in most cases—just not pruned excessively more often than needed. Prune your plants for a good reason, and you will enjoy increased beauty with less work and saved water!

## Don't Let Weeds Steal Water

Weeds—we usually don't want them in our gardens. There are native weeds and nonnatives present in the garden, but both are often unwelcome guests. Usually, the reasons we dislike weeds range from their appearance, how they spread so quickly, and the allergies some of them cause. However, some weeds provide benefits in the garden, such as dandelions that are edible and provide flowers for pollinators and clover, which helps increase the fertility of the soil. However, many weeds cause problems in the garden setting. Weeds are very good at stealing water from your plants. They naturally come up where moisture is present, so weeds frequently pop up near your plants, outcompeting them for available water. Often, weeds can be sneaky, taking root underneath plants where they hide, grow, and use water meant for desired plants.

### How to Control Weeds

Keeping weeds away from your plants is essential as they steal water and nutrients. Often, the first impulse for many people is to grab a bottle of commercial weedkiller (herbicide). However, the use of these herbicides is controversial, and they aren't suitable for killing weeds next to plants without injuring the plant itself. Thankfully, it is possible to

A prickly lettuce (*Lactuca serriola*) weed is stealing water and nutrients from a groundcover.

manage weeds using alternative methods without herbicides. Mulch will help to minimize unwelcome weeds around plants. Periodically look under a plant's foliage to ensure no weeds are lurking there. Remove weeds as soon as you notice them. Pulling weeds is easier after a rainstorm when the soil is wetter. Hula or stirrup hoes are useful weeding tools. They are a long-handled modified hoe with a sharp blade that cuts off the tops of weeds at the soil surface.

#### *Don't Use Plastic for Weed Control*

Whatever type of mulch you choose, avoid adding sheets of landscape plastic underneath your mulch layer to prevent weeds. The use of plastic to reduce weeds has several adverse effects on the garden. First, it prevents rain from reaching the roots of plants and, as a result, increases the amount of supplemental water plants require. The plastic also causes low levels of oxygen, and the soil underneath has a hard time drying out, causing problems for plants with overly wet soil. Over time, the plastic becomes exposed and breaks apart into smaller pieces, which is unsightly. Weeds still can make their way into gardens with landscape plastic. Skip the plastic in the garden!

## Fertilizing Practices for Water-Smart Gardens

When we think about how to care for gardens, pruning, watering, weeding, and fertilizing come to mind. However, many people tend to do more for their plants than they need, which can increase how much water the plants need. This also extends to applying fertilizer to plants when it isn't needed. Fertilizers supply nutrients for plants to make up for what existing nutrients in the soil don't provide. The role of fertilizer's effects on plants' water needs is influenced by the kind of fertilizer, the type of plant, and whether the plant needs extra nutrients.

One-size-fits-all doesn't work for plants regarding fertilizer use—some plants have higher nutrient requirements, while others are perfectly happy with lower levels. No supplemental fertilizer is needed if your soil has sufficient nutrients for a plant's needs. Plants native to your region are usually content with nutrient levels already present in the soil. However, plants from areas with higher nutrient soil than yours will likely need fertilizer. Unfortunately, many gardeners apply fertilizer to plants regardless of whether they need it. When plants have more nutrients than they need, it can lead to a flush of green growth at the expense of flowers, or straggly or floppy stems or branches. Fertilizing plants when they don't need it stimulates growth, increasing their water needs.

### Does My Plant Need Fertilizer?

Some plants have high-nutrient needs and require regular fertilizer applications, such as container plants, flowering annuals, fruit and vegetable gardens, lawns, and roses. Native plants, such as trees, shrubs, and perennials, or those from regions with similar soil nutrient levels, likely won't need any supplemental fertilizer. Or, a single application of a slow-release fertilizer in spring will fulfill their needs. However, if you have plants from regions known for nutrient-rich soil—higher than yours—they will likely need to be fertilized. When in doubt, don't fertilize unless you see signs of nutrient deficiency, such as yellowing leaves or yellow leaves with dark green veins.

### What Type of Fertilizer Is Best for Water Efficiency

The kind of fertilizer you use also affects the water usage of plants. Inorganic (chemical) fertilizers are made from salts, which require more water to allow the fertilizer to be absorbed into plant roots. Additionally, inorganic fertilizers tend to have high amounts of salts, which can increase the salinity of your soil, which isn't good for your plants. Organic fertilizers such as compost, aged manures, or plant-based fertilizers are the more waterwise option as they have lower salt levels than chemical fertilizers. They also help improve soil texture, contributing to a healthy water-holding capability.

## SOIL TESTING

If you grow various plants from different regions, it's helpful to know the nutrient levels of your soil to determine whether you need to fertilize to provide additional nutrients. For a soil test, soil samples are taken from different areas around your garden, mixed together, and sent to a soil laboratory. Test results show the levels of nutrients in your soil, the level of organic matter (a higher percentage means more fertility), and your soil pH. You can use this information to provide needed nutrients.

Shrubs showing signs of iron chlorosis, due to overwatering.

## Best Low-Water Practices for Lawns

Lawns aren't low-water users and aren't recommended for dry climate regions. However, lawns are more commonplace in regions where rainfall is usually plentiful. How you maintain your lawn affects its water usage, especially when it comes to mowing height and frequency. Frequent mowing increases the water needs of grass as it stimulates rapid regrowth. Allow your grass to grow taller to 2 inches (5 cm) for warm-season grasses and 3 inches (8 cm) for cool-season grasses. The taller grass height also helps to decrease weed growth. The increased height helps shade the grass's roots and prevents the soil from drying out quickly. To determine how often to mow your lawn, mow when it hits the recommended height. Cut no more than one-third of the height of the grass each time to prevent water stress that can occur when cut too short. Finally, aerate your lawn every few years during the growing season to decrease compaction and allow water to permeate more deeply into the soil. See chapter 8 for low-water lawn alternatives.

In warmer climates, where Bermuda grass turns dormant in winter, a frequent practice is to overseed the grass with perennial ryegrass seed to ensure lush green color through the winter months. A lot of water is initially required for the overseeding process to help the seed sprout, and then regular watering is needed to maintain the grass through the cool season. One simple step to help save water is to ditch the overseeding and allow your grass to stay dormant through the winter, which requires much less water and results in a lawn that can typically exist on natural rainfall until spring. A bonus to skipping overseeding is healthier Bermuda grass, as repeated seasons of overseeding can cause competition and bare spots amidst the Bermuda grass.

Another water-saving lawn maintenance task is to recycle grass clippings. When you mow your lawn, don't bag the grass clippings—instead, allow them to fall back onto the turf. The grass clippings break down very quickly, releasing nitrogen into the soil and reducing the need for supplemental fertilizer. This practice saves you time and labor and helps build more permeable soil so water can penetrate more efficiently, decreasing the amount of supplemental water needed.

Adjust your mower to a higher height to allow grass to grow taller, which will shade the roots, decreasing water use.

## Other Water-Saving Tasks

In addition to mulching, weeding, and proper fertilizer usage, we must incorporate two other maintenance tasks into our garden calendar. If you use organic mulch, it's essential to reapply a fresh layer every year in spring or fall. Organic mulches break down over time, enriching the soil, and reapplication is vital to enjoying all its benefits. The second task is to make seasonal adjustments to your irrigation schedule. Plants' water needs go up and down in response to heat and day length—more frequent irrigation is needed in summer than in fall and spring. There may be little to no need to provide supplemental irrigation in winter.

CHAPTER 8

# Replacing the Thirsty Lawn

**The most significant water savings** in the landscape come from removing or reducing the size of the lawn. In the United States alone, grass lawns require more water more often than most other plants, which is true of many other regions worldwide. Drive through almost any neighborhood, and you will likely see front yards filled with a lawn, even in drier climates, despite the large amount of supplemental water required to maintain them.

A backyard lawn with weeds intermingling with the grass that require frequent maintenance to remove them.

## Why Are Lawns So Popular?

You may be surprised to learn that residential lawns are a recent trend in the garden. The possibility of having a lawn didn't exist for the average homeowner until the late 1800s. Before that time, gardens were filled with flowering plants and edible plants such as vegetables and herbs. In the early twentieth century, grasses suitable for a lawn were introduced, such as warm-season Bermuda grass from Africa, blue grasses from Europe, and fescues from North America. At the same time, the invention of the lawn mower and water being piped directly to homes made it possible for many homeowners to have a manicured lawn.

However, the ability to have a lawn comes with negative environmental impacts from water use, pesticides, chemical fertilizers, and carbon emissions from lawn equipment. Homeowners tend to use more pesticides and fertilizer on their lawns than required, which can find their way into water supplies due to runoff. People often want a lawn for their kids or pets to enjoy; however, improper or overuse of fertilizer and pesticides to maintain the lawn can expose children and pets to unwanted substances.

## How Much Water Do Lawns Use?

The average lawn requires approximately 1 inch (3 cm) of water per week during the growing season. You may seldom need to water your lawn in climates with plentiful rainfall. However, if you live in a dry climate, experience drought, or have low rainfall seasons, supplemental irrigation is needed to grow a healthy lawn. A square foot of lawn requires 0.623 gallons (2.36 L) of water to reach the recommended 1 inch (3 cm) weekly. Suppose you have a 1,000-square-foot lawn (1,000 sq ft × 0.623 gallons [93 sq m × 2.36 L])—that equals 32,396 gallons (122,616 L) of water yearly just for your lawn. Your lawn may be larger or smaller than that, but you can make a significant difference in your water usage by replacing the lawn with other types of plants.

The good news is that you can have a beautiful garden filled with flowering plants, lush green groundcovers, and other types of plants that use a fraction of the water a lawn does. Another bonus is less maintenance, pesticides, and fertilizers will be needed. You also will save money with a reduced water bill, not having to pay someone to mow your grass, and needing less or no fertilizer or pesticides.

### Why You Should Consider Getting Rid of Your Lawn

In dry climates and drought-prone regions, lawns aren't a sustainable landscape practice. Yet, they were (and still are) a popular feature of many residential landscapes. Water restrictions instituted in response to drought in many cities make it difficult or impossible to supply the water needs for a lawn. As a result, we see many homeowners with dead lawns waiting for water restrictions to be lifted so they can grow them again. In some places, where water supplies are low, lawns are no longer allowed, and existing ones are being ripped out. The good news is that many municipalities are offering rebates for lawn removal to encourage residents to convert to a low-water landscape.

#### *When Lawn Removal Isn't Possible*

There is a lot of controversy surrounding lawns, particularly in dry climate regions where water supplies are shrinking and lawns require a lot of supplemental water. However, for some people, lawn removal isn't an option, which can be due to several factors, including neighborhoods that require a lawn or a spouse (or partner) who can't bear parting with their patch of green lawn. If this describes your situation, it's essential to exercise waterwise strategies to use water efficiently to keep your grass healthy (see chapter 7, page 117) for recommended lawn maintenance practices.

## What Should I Do If I Remove My Lawn?

Perhaps your desire to remove your lawn is motivated by being tired of the high maintenance—from frequent mowing, weed control, and fertilizing. Weeds always try to infiltrate our lawns and require constant attention to manage them. It may be that your lawn doesn't look as nice as your neighbor's, who spends hours keeping it perfect. The "perfect" patch of grass takes a lot of work.

It can be hard to consider taking out your lawn—people love their swath of green. Thankfully, there are options for you to consider to significantly decrease the water usage and maintenance of a traditional lawn.

- Consider taking out nonfunctional areas of grass that only have decorative value, such as that patch of green in the front yard. See chapter 11, page 158 on how to remove your lawn.

A lawn made of white clover needs less water, mowing, and fertilizer compared to a traditional lawn.

Lawn watering in the middle of the day causes up to 50 percent of water to evaporate instantly.

Before and after photos of a front yard landscape converted to a xeriscape filled with 'Blue Elf' Aloe (*Aloe* × 'Blue Elf'), purple trailing lantana (*Lantana montevidensis*), firecracker penstemon (*Penstemon eatonii*), and angelita daisy (*Tetraneuris acaulis*).

- Reduce the size of your lawn, which will help to save water. Focus on removing small strips of lawn or areas that extend beyond your view. Consolidate your lawn into a centralized area where it will likely be used, and maintenance is more manageable.
- Another option is to replace your grass lawn with a lawn alternative that can handle light foot traffic. White clover (*Trifolium spp.*), creeping thyme (*Thymus serpyllum*), and Kurapia (*Phyla nodiflora* 'Kurapia') are popular lawn substitutes to explore. These lawn substitutes also produce flowers, which attract bees.
- Replace your lawn with a specialized turfgrass mixture made up of native grasses that use less water and need less maintenance yet resemble the appearance of a lawn. Look for a specialized turfgrass mixture for your region.
- Grow wildflowers as a substitute for a lawn, adding beauty when in bloom and attracting pollinators.
- Utilize the green spaces in your community for you and your family.

## Save Water with a Beautiful Xeriscape Garden

A popular style of low-water landscape is xeriscape, which entails using low-water methods and plants to create an attractive landscape. A xeriscape uses half to two-thirds less water than one with a lawn. Xeriscaping was created in 1981 in Denver, Colorado, to encourage people to ditch their thirsty lawns for an attractive, low-water-use garden. Many people incorrectly refer to this style as "zero-scape," which is misleading because a well-designed xeriscape is a sustainable, plant-filled oasis. This method isn't just for dry climates but is implemented throughout all gardening regions. There are seven principles to creating a xeriscape:

1. **Planning and design** Evaluate your site for topography, sun exposure, and soil type. Create a preliminary design—a simple sketch will do. Make a wish list of the elements you want in your outdoor space, such as an entertainment area or a quiet space to relax. Also consider plants for color, shade, attracting wildlife, or vegetable gardening. Select plants according to the existing conditions of your site, matching them with sun exposure and soil type. Be sure that they have enough room to grow to their mature size. Group plants with similar water needs together. See chapter 3 on how to place plants for water efficiency.
2. **Choose low-water plants** Utilize native and other plants adapted to the rainfall, temperature, and humidity levels in your climate. Plants that are considered drought-tolerant in your region are a great choice. Ensure plants can handle environmental extremes within your climate. Visit your local botanic garden for plant inspiration. See chapter 2 for plant selection tips.
3. **Improve soil** Amend clay or sandy soils for nonnative plants to improve water capacity. Add compost to sandy soils to allow them to hold onto water longer. For clay soil, incorporate compost to help improve drainage. Native plants seldom require soil improvements. For tips on how to improve soil, see chapter 4.
4. **Efficient irrigation** Utilize water-efficient irrigation methods such as drip irrigation for trees, shrubs, and groundcovers. Ensure your irrigation system is free of leaks and operating efficiently. Water early in the morning to decrease evaporation. Apply water long enough to permeate deeply to encourage deep root growth. Adjust watering frequency seasonally (consult chapter 6 for water-efficient guidelines). Incorporate rainwater-harvesting strategies within the garden.

A newly planted landscape in Phoenix, Arizona, planted with Outback Sunrise Emu (*Eremophila glabra* 'Mingenew Gold') groundcover.

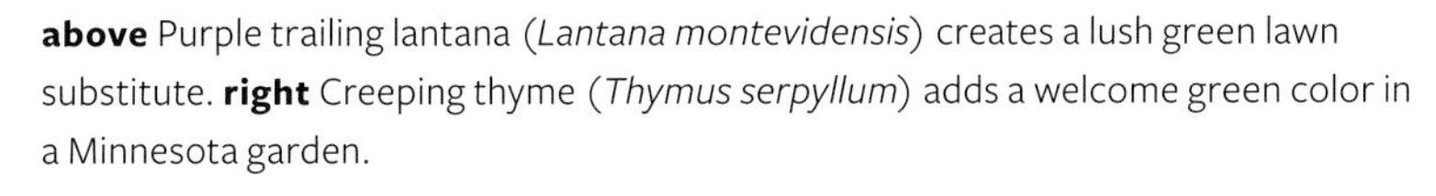

**above** Purple trailing lantana (*Lantana montevidensis*) creates a lush green lawn substitute. **right** Creeping thyme (*Thymus serpyllum*) adds a welcome green color in a Minnesota garden.

5. **Use mulch** Cover bare soil with a 2-inch (5 cm) layer of organic or inorganic mulch to prevent soil from drying out too quickly around plants. Wood chips and pine needles are examples of organic mulch, while decomposed granite and landscape rock are inorganic mulches.
6. **Reduce or eliminate grass areas** Eliminate all nonfunctional turfgrass and minimize or replace functional grass areas with low-water alternatives (explore turf removal options in chapter 11, page 158). Replace lawn areas with low-growing groundcovers, flowering perennials, or shrubs that provide beauty and habitat.
7. **Appropriate maintenance** Practice proper garden tasks according to plants' needs. Avoid overwatering, fertilizing when not needed, and excessive pruning, which increases a plant's water needs.

## Groundcovers Instead of Grass

If you are passionate about having a carpet of green in your garden but don't want artificial turf and want to save water, consider using grasses native to your region. Or, how about grouping waterwise groundcovers with a similar shade of green that a lawn has? While most groundcovers may not be walkable, they will add the attractive splash of green color a lawn does. The key to creating an attractive swath of green with groundcovers is to space them so they will grow together, covering bare areas in between. Look for groundcovers that stay below 2 feet (61 cm) in height. There are many lawn replacement options for you to explore that will bring beauty and water savings to your garden.

### DROUGHT-TOLERANT GROUNDCOVER ALTERNATIVES FOR GRASS LAWNS

- Wood's Sedge (*Carex woodii*)
- 'Green Carpet' Natal Plum (*Carissa macrocarpa* 'Green Carpet')
- 'Outback Sunrise' Emu (*Eremophila glabra* 'Mingenew Gold')
- Creeping Juniper (*Juniperus horizontalis*)
- Trailing Lantana (*Lantana montevidensis*)
- Sunshine Mimosa (*Mimosa strigillosa*)
- Creeping Phlox (*Phlox stolonifera*)
- Dwarf Elephant Food (*Portulacaria afra minima*)
- Dwarf Cinquefoil (*Potentilla canadensis*)
- Carpet Sedum (*Sedum lineare*)
- Wedelia (*Sphagneticola trilobata*)
- Creeping Thyme (*Thymus serpyllum*)

*Not all of these groundcovers can tolerate foot traffic or will grow where you live. Consult with your local nursery professional to see which one of these groundcovers is suitable for your climate and needs or to help provide additional groundcover options.

## What about Artificial Turf?

Many people are replacing their lawns with artificial grass so they can enjoy the rich green color year-round. This grass substitute rolls out like a green carpet onto the ground. It is made from plastics and top-dressed with a ground-up rubber or sand layer to help the grass stand upright. However, before you decide to add a synthetic lawn to your landscape, it's important to consider the facts about artificial turf to help you make the right decision for your needs:

- It is expensive to install.
- The hot surface makes it uncomfortable to walk or play on in summer.
- Plant debris collects on it until blown off or vacuumed off. Periodic spraying of the turf with water is needed to remove dust.
- Weeds can grow in artificial turf.
- Color can fade over time.
- Pet waste is harder to remove, and odors can build up, so periodic disinfection is needed.
- It requires periodic "brushing" to keep it upright.
- Water and oxygen cannot permeate the soil beneath, hindering soil health and nearby plant roots.
- Plastic and rubber infill chemicals may cause adverse health effects, particularly for young children and pets, which is currently being studied.

If you have a lawn, I invite you to reimagine your landscape without a turf lawn, replaced with one filled with plants that add beauty, attract pollinators, use much less water, and are much easier to care for.

Artificial grass collects leaf debris that needs to be removed by blowing or vacuuming. Water doesn't permeate the fake grass, so the tree has trouble getting the water it needs.

CHAPTER 9

# Waterwise Container Gardening

**Container gardening is a great way** to grow plants when space is limited, to spruce up a bare corner, or to increase your curb appeal. When most people think of growing plants in a pot, flowering annuals come to mind, and while they add a welcome splash of color, they require frequent watering, especially during the summer months. However, I invite you to expand your view of what can grow in a container. Simply put, a container is a vessel that contains a small garden, and the options for what you can plant in it are almost limitless.

The good news is that containers work well with waterwise plants that need less water, fertilizer, and maintenance than flowering annuals, and they look beautiful! Additionally, you don't have to purchase new plants every season, which saves you money. Keep in mind that growing a plant in a container will require more water than if grown in the ground because the soil in pots dries out faster. However, there are ways to enjoy container gardening with water efficiency in mind. Let's begin by exploring five strategies: location, container type, soil, plant choice, and watering.

Shrubs, Blue Bells™ emu (*Eremophila hygrophana*), and Mexican honeysuckle (*Justicia spicigera*) make a colorful statement in the author's garden while using less water than flowering annuals.

## Container Garden Location

Flowering lantana and rosemary are an excellent alternative to thirsty flowering annuals.

Colorful, glazed containers are a great way to add vibrant interest to the garden when using low-water plants.

Before we select our pots and plants, it's essential to identify where we will place them, as their location affects what type of plants we choose and the type of container. Where you put your containers also affects how well the plants will do and how much water they will require. Be sure to note a plant's preferred exposure and place them accordingly. Full sun exposure all day, or in the afternoon, is stressful for plants growing in a container and increases water usage. Therefore, it's best to place potted plants in areas that receive afternoon shade or filtered sunlight when the day's heat is at its highest. You also can move containers throughout the year to ensure the ideal sun exposure, especially if they are on a base with wheels. Grouping several planted containers also can help decrease water usage by creating a mini microclimate that increases humidity and decreases the temperature by a few degrees. A bonus is that it also makes watering and upkeep easier when they are near each other. Consider adding a wheeled base on the bottom of your containers to allow you to move them easily to a different sun exposure in response to seasonal changes.

Following these tips in planning your container garden will save water while also allowing you to enjoy healthy and thriving plants. Let's discuss plant choices, watering, maintenance, and additional water-saving tips.

## Container Type Matters

Walk into any nursery, and you will encounter a variety of container choices, yet some containers make better choices than others for waterwise container gardening. Glazed or plastic pots work well for minimizing water use. While corten steel or galvanized steel planters are popular, they aren't the best choice for climates with hot summers, as hot air temperatures can "cook" the roots of the plants as the steel gets very hot. Containers made from wood, such as half whiskey barrels, work well but have a shorter lifespan than glazed, plastic, or steel pots.

Whatever your container choice, ensure there is a hole for drainage because soggy soil will kill plants. The size of your pot matters—so select a larger container. With a small container, the soil can only hold onto a little moisture and dries out quickly, requiring frequent watering. If you've ever struggled to grow plants in a small container in the summer, a combination of heat and dried-out soil is likely the culprit. A larger-sized container holds more soil and water, so you'll need to water less often. Avoid using dark-colored containers in sunny areas, which heat up more and cause plants to use more water. Start at a minimum size of 20 inches (50.8 cm) tall and wide for perennials, shrubs, or larger succulents; if you are growing a large shrub or small tree, you will need an even larger pot.

In climates that experience hot summers, root stress is common for plants in the container garden, which increases their water usage. Root stress is especially prevalent for plants growing in full sun in smaller pots with little to no insulation from the heat for the roots. You can provide insulation for the roots by double potting your plants. Simply keep your plant in the nursery container and place it in a larger pot. Another option is to place a patio chair or other object to block the sun's rays from hitting the container during hot summer weather.

### RETHINK TERRACOTTA CONTAINERS FOR LOW-WATER GARDENS

Avoid using terracotta containers for leafy green plants such as flowering annuals, perennials, or shrubs. Terracotta is a porous material that allows water to evaporate through the sides of the pot, so more water will be required to keep plants hydrated. Terracotta is okay for succulents, which prefer dry soil in between watering.

### Self-Watering Containers

Self-watering containers are an increasingly popular option for those who want the versatility of container gardening without having to water plants frequently. Another bonus of self-watering containers is that they take away the risk of overwatering plants, a common mistake. These modified containers have a reservoir at the pot's base, which is filled with water. The potting soil absorbs water from the reservoir as the soil starts to dry, which ensures a consistent level of moisture

Higher water-need plants such as annual flowers and vegetables require less frequent hand watering when planted in a self-watering container.

that leads to healthy plants due to no dry or overly wet soil. The reservoir must be filled with water every few days as the level drops. Self-watering containers allow for more freedom if you can't provide water for plants frequently or if your pot is in an inconvenient location for water access. When shopping for a self-watering container, ensure it has an overflow hole for excess water to flow out to prevent soggy soil.

If your containers depend on hand-watering, then self-watering containers are an excellent option in most cases. They also help to conserve water since water is absorbed from the bottom of the soil and not the top, which decreases evaporation. Additionally, it helps prevent fungal disease that can occur when watering plants from the top.

Self-watering containers aren't suitable for succulents, which do best when soil completely dries out for several days between watering. Succulents will rot in soil that is continually moist. Use traditional planters or terracotta pots for growing succulents.

## Getting the Right Soil Mix

Walk into any nursery, and you will encounter a myriad of bagged soil mixes. Understandably, you may feel overwhelmed about choosing one that will work best for your needs. Don't worry; it's easier than you think to select the right one for you. Regular potting soil isn't always the best choice for containers because it can hold onto too much water, which isn't suitable for plants—especially drought-tolerant plants. Look for a product that lists "mix" in the title, such as "potting or planting mix," or one that lists "container" in its description. These soil mixes are created to hold the right amount of moisture for container plantings without becoming too wet and work well for growing all types of plants except those requiring well-drained soil, such as succulents.

## Waterwise Container Soil Amendments

Once you have selected your container soil mix, consider adding amendments to your potting mix that help with water retention and allow more time between having to water your pots. Vermiculite is a nontoxic, natural material that absorbs water and releases it as the soil dries out and doesn't break down over time. The amount of vermiculite in your soil mix should be approximately 25 percent. Another amendment that helps with water retention is coco coir, which is ground-up coconut husks. It is very good at absorbing water and can be incorporated into container soil so that it makes up 30 percent of the total mixture. Coco coir will break down over time but lasts up to four years.

Other amendments to add to your potting mix include worm castings and compost to create healthy soil for growing perennials and shrubs when added annually in spring. Plants with high nutrient requirements, such as flowering annuals and vegetables, will need supplemental fertilizer applications (see page 135).

### *Water-Absorbing Beads*

Many container soil mixes contain small beads of gel made of a water-absorbent polymer. These beads absorb and hold a substantial volume of water. As the soil dries out, the beads release water into the soil. Using these polymer gels helps provide an even level of moisture and leads to less frequent watering. However, they break down over time, especially in hot climates where exposure to the sun accelerates their degradation. There are also some questions about possible adverse environmental and health risks, with further studies being performed. Coco coir or vermiculite are excellent alternatives to water-absorbent beads.

## Succulent Soil Mix

If you grow succulents in a pot, a regular container potting mix isn't suitable for their needs. Cacti and other succulents must have fast-draining soil—remember that they prefer dry soil between watering cycles, so they need less frequent watering than other plants. As a result, we need to use a soil mix that drains quickly, where the roots of succulents can absorb water when watered and then enjoy a dry respite. Succulent soil mixes don't hold onto water for long and don't provide consistent moisture, which is harmful to succulent plants. You can purchase ready-made container soil mixes formulated for succulents, or you can make your own.

### DIY SUCCULENT PLANTING MIX

You can save money by making your own soil mix for succulents, and it's easy to do. You will need three ingredients:

- Perlite
- Coarse builders sand
- Potting soil

Your soil mix should consist of one part perlite, two parts coarse builders sand, and three parts potting soil combined.

A collection of agave and aloe succulents planted in terracotta containers add sculptural interest in a Texas garden.

## Waterwise Plant Options for Containers

Flowering annuals are the most popular option for containers; however, they are the most water-intensive, requiring frequent water applications. Vegetables are also high-water-use. The good news is that the same drought-tolerant plants in your garden are likely suitable candidates for growing in containers. When shopping for container plants, keep in mind that they will grow smaller than when planted in the ground because there is less space to grow roots. You can add multiple plants per pot (if there is enough room) or plant a single one. If you grow multiple plants in a container together, ensure they have similar water requirements.

### Flowering Shrubs and Perennials

You may love the beauty of flowering annuals but not how much they need water and frequent maintenance, such as fertilizing and deadheading. Also, you need to replace them seasonally. Years ago, I decided to switch out my thirsty annuals for colorful, flowering shrubs and perennials, and I have never regretted my choice. First, they need much less care and water than annuals, and I still get the beauty and color I desire. Shrubs can live for years in a container if properly cared for, and perennials are usually suitable for two to three years. Pollinators are frequent visitors to my flowering container plantings, which makes them even more enjoyable. While container shrubs and perennials need water more often than those planted in the ground, they need much less water than annual flowers. Containers should be at least 20 inches (50.8 cm) tall and wide at the top for medium to large-sized shrubs.

If you live in an area with cold winters, consider using perennials as annuals to add welcome color without the high water needs of annuals. Note that plants grown in pots outdoors are subject to more cold stress, so use plants that can tolerate temperatures at least 10°F (6°C) lower than you usually experience in winter.

A heavenly bamboo (*Nandina domestica*) shrub makes an attractive container plant that needs less water and maintenance than flowering annuals.

Most shrubs and perennials are suitable for growing in containers, and some cacti and succulents work better than others for container gardening. Have fun designing your containers. Play with different color and texture combinations. If you mix plants together in the same container, make sure they share the same sun and water requirements. Consult with a nursery or garden professional for suitable plant choices for your climate, container size, and location in the garden.

A wonderful alternative to flowering annuals are herbs, which do great in containers and use less water. Herbs have attractive foliage, which adds a decorative element, and they are steps away from your kitchen as a source of delicious flavor for your favorite dishes. You can plant different herbs together or singly.

A row of potted succulents thrive along a sunny window ledge during the summer months in Minnesota.

### Succulent Container Gardens

A popular option for a very low-water container garden is using cacti and succulents (cacti are a type of succulent). Succulents are plants that can store water and use it when water isn't available. Their unusual shapes also add a unique accent to your outdoor space. If you have an area in your garden with limited access to water, grow succulents—you won't have to water them too often, and you'll still enjoy an attractive container planting. One crucial aspect of growing succulents in a container is that the container soil must be fast-draining—they will die if their roots remain moist for too long. People who struggle to grow succulents usually overwater them, and it's easy to fall into this trap—we want to "baby" our plants, but often it is with negative results. *You must allow the soil to completely dry out for a week (or more) before watering again* (see container watering guidelines, chapter 8, page 136).

You can group several succulents in a single container or grow one as a statement plant. However, don't plant succulents together with annuals, shrubs, or perennials, which require more water—this will likely lead to the succulent struggling to survive, as too much water isn't good for it.

## Container Plant Care

Plants grown in containers require more maintenance than those grown in the ground when it comes to fertilizing, pruning, and watering. Any plant grown in a container will require supplemental fertilizer because they have no access to nutrients in the ground. More frequent pruning also is needed to keep plants from outgrowing their pot. The soil in containers dries out more quickly, so plants need water more often. Let's break down each maintenance task.

## Lower Water Plant Options for Containers

| | |
|---|---|
| **GROUNDCOVERS/ PERENNIALS** | Yarrow (*Achillea* spp.), artemesia (*Artemisia absinthium*), coreopsis (*Coreopsis* spp.), blanketflower (*Gaillardia aristata*), verbena (*Glandularia* spp.), sweet potato vine (*Ipomea batatas*), trailing lantana (*Lantana montevidensis*), lavender (*Lavandula* spp.), gaura (*Oenothera lindheimeri*), penstemon (*Penstemon* spp.), mealy cup sage (*Salvia farinacea*), Russian sage (*Salvia yangii* syn. *Perovskia atriplicifolia*), lamb's ear (*Stachys byzantina*), and purple heart (*Tradescantia pallida*) |
| **SHRUBS** | Bougainvillea (*Bougainvillea* spp.), boxwood (*Buxus* spp.), hop bush (*Dodonaea viscosa*), euonymus (*Euonymus* spp.), hydrangea (*Hydrangea* spp.), Arabian jasmine (*Jasminum sambac*), Mexican honeysuckle (*Justicia spicigera*), lantana (*Lantana* spp.), Texas ranger (*Leucophyllum* spp.), blue potato bush (*Lycianthes rantonnetii*), dwarf myrtle (*Myrtus communis* 'Compacta'), heavenly bamboo (*Nandina domestica*), and yellow bells (*Tecoma* spp.) |
| **ORNAMENTAL GRASSES** | Carex (*Carex* spp.), dwarf pampas grass (*Cortaderia selloana* 'Pumila'), Blue fescue (*Festuca glauca*), pink muhly grass (*Muhlenbergia capillaris*), deer grass (*Muhlenbergia rigens*), and Mexican feather grass (*Nassella tenuissima*) |
| **SUCCULENTS** | Agave (*Agave* spp.), aloe (*Aloe* spp.), desert spoon (*Dasylirion wheeleri*), ice plant (*Delosperma* spp.), sansevieria (*Dracaena trifasciata* syn. *Sansevieria trifasciata*), euphorbia (*Euphorbia* spp.), red yucca (*Hesperaloe parviflora*), golden barrel (*Kroenleinia* syn. *Echinocactus grusonii*), hardy spineless prickly pear (*Opuntia cacanapa* 'Ellisiana'), totem pole (*Pachycereus schottii f. monstrosus*), elephant food (*Portulacaria afra*), sedum (*Sedum* spp.), and yucca (*Yucca* spp.) |

Consult with a garden professional in your area to determine which plants from this list are suitable for containers for your climate.

### Fertilizing

A slow-release fertilizer is a good choice for plants such as shrubs, perennials, and succulents, as it slowly releases nutrients throughout the growing season. Follow the fertilizer label's directions on how much to apply and how frequently. For flowering annuals or vegetables, apply a water-soluble fertilizer every few weeks. Cacti and other succulents require less fertilizer than shrubs and perennials. Fertilizing them once in spring with a slow-release fertilizer once the threat of frost is over is usually sufficient for their needs.

### Watering Containers the Right Way

Any plant grown in a pot will dry out faster than if planted in the ground. Each time you water your container garden, you must water deeply to ensure the entire root ball is wet and water drains from the bottom. If your soil has become completely dry, water can have difficulty permeating soil and run off the side of the soil instead. To rectify this, take a screwdriver and poke several holes through the soil to allow water to soak in. Look for water to drain out the hole. If the soil isn't completely moist after watering once, wait an hour and water again.

The most common question regarding container gardening is, "How often do I need to water my plants?" The answer is dependent on several factors. To start, the size of the container matters—larger pots hold onto more water and need less frequent watering than smaller ones. The time of year also matters—in summer, the soil around plants will dry out more quickly than in spring, fall, or winter.

Where you place your pots also affects how much water is needed—those in full sun need more water than those in shade. Wind and humidity levels also influence how often a plant needs water. Finally, different types of plants have varying water needs. All these factors determine your watering frequency. Except for succulents, most container-grown plants want consistently moist (but not overly wet) soil.

Monitor your plants for signs of drought stress, such as wilting leaves or succulents showing evidence of wrinkling stems or leaves. Ideally, we don't want our plants to get to the point where they are showing stress from too little water. Moisture meters are a common tool for determining when to water your potted plants. However, they aren't always accurate. Soil type, temperature, and salinity can provide misleading results. However, another test provides more accurate results—your finger.

### *Efficient Watering Methods for Container Gardens*

Plants can be hand-watered using a watering can or a "rain wand," which gently applies water without disturbing the soil like a hose can. Less labor-intensive watering options include using a self-watering container or placing an olla in the soil among the plants. Ollas provide even watering and allow you more time between irrigating your plants. The olla is filled with water and slowly releases it into the surrounding soil (see chapter 6, page 102). Note that ollas aren't suitable for succulent plants, which need periods of dry soil.

The easiest and most worry-free way to water containers is to attach them to a drip irrigation system to water them on a schedule if a timer is attached. This is particularly helpful if you are gone for extended periods. Drip irrigation can be threaded through the drainage hole before adding soil or from the sides of existing container gardens.

## USE THE "FINGER TEST" TO TELL WHEN TO WATER

The finger test consists of sticking your index finger into the soil as far as it will go. Bring it back up and rub it against your thumb to determine the soil moisture. If it feels wet or moist, soil particles will stick to your finger, and you don't need to water that day. However, it's time to water if your finger is mostly clean and the soil feels mostly dry. The exception to the finger test are succulents, which like dry soil for at least a week, or two between watering cycles.

## Add Mulch to Containers

Mulch is a great way to prevent soil from drying too quickly in the garden, and this is also true for container plantings. After planting, add a 2-inch (5 cm) layer of organic mulch, such as wood chips, around your shrubs or perennials, and reapply annually. Use a landscape rock such as pea gravel for a neat and clean look for cacti and other succulents.

Water container plants at the base to prevent wetting foliage and encouraging fungal disease.

CHAPTER 10

# The Water-Efficient Vegetable Garden

**Growing your own produce is immensely rewarding,** and nothing can beat the delicious taste of home-grown vegetables. There is also the joy you experience when you eat food you have grown yourself. However, vegetable gardening can use a lot of water since most vegetables aren't low-water plants. Historically, vegetables have been grown in wet and dry climates, so to ensure we are using water efficiently, we can implement old and new techniques to ensure we make the most of our water use in the vegetable garden.

A vegetable garden in Phoenix, Arizona, with warm-season crops such as beans, corn, and squash, along with herbs growing in a container.

## Vegetables Aren't Low-Water

Most vegetables are annuals and complete their life cycle within a single year or growing season. This means they grow quickly to produce their fruit when seasonal conditions are just right. In general, fast-growing, annual plants are higher water use. Some vegetables are tender perennials, which can live longer than a year but are usually treated as annuals.

Vegetables are put into two groups: cool season and warm season. Cool-season vegetables are grown before and after the summer season in cooler climates, while in regions with mild winters, they can grow through the winter months. Vegetables grown during cooler times of the year use less water since temperatures are more moderate and the sun is not intense. Warm-season vegetables are grown through the summer in temperate climates and in late spring to early summer in regions with hot summers. Unsurprisingly, warm-season vegetables use more water due to higher temperatures and intense sunlight. The timing for planting cool- and warm-season vegetables is based on the region where you live and can be accessed online or by consulting with a garden professional in your area.

### Cool-Season Vegetables

Beets, broccoli, cabbage, carrots, cauliflower, garlic, kale, lettuce, onion, peas, radish, leafy greens, peas, potatoes, Swiss chard, and turnips

### Warm-Season Vegetables

Beans, cantaloupe, corn, cucumber, eggplant, okra, peppers, pumpkin, squash, sweet potato, tomatoes, watermelon, and zucchini

## Techniques for Decreasing Water Use In the Vegetable Garden

You can save water in a variety of ways while growing vegetables. Whether you have an existing plot of vegetables or are planning to add an edible garden, you can implement water-saving tips while experiencing the joy of growing produce. From the structure of your vegetable garden, to how you water it, to vegetable selection, we can use methods to manage our water use wisely.

### Vegetable Garden Location for Water Efficiency

Location matters when it comes to growing vegetables successfully, but it also affects how much water your vegetable garden requires. Most vegetables do best when they receive six to eight hours of sun (if you live in a desert region with intense sun, six hours is usually sufficient). Like most plants, vegetables need more water in the hottest part of the day, typically mid-afternoon. For regions that experience high heat in summer, select an area that receives shade by mid-afternoon, which decreases the temperature while minimizing the sun's influence. If there is no afternoon shade available where your vegetable garden is, you can provide it during the summer months using shade cloth (see chapter 3, page 50).

Place the garden near a water source, such as an outdoor faucet, for easy watering. Water is heavy, and carrying it from a long distance is difficult and not very sustainable over the long term due to the frequent watering required by vegetables.

Avoid locating your garden near a tree. Plant roots grow wherever there is water, and those larger plants will grow into your garden, stealing

Vegetable garden planted in the ground to minimize the soil from drying out too quickly.

A shallow, raised vegetable bed that minimizes soil drying versus a taller raised vegetable garden.

water from the vegetables. Locate the vegetable garden at least 10 feet (3 m) from small- to medium-sized trees and even farther from large trees. If you have large shrubs over 5 feet (1.5 m) tall, keep the garden at least 7 feet (2.1 m) away to avoid root encroachment.

## In-Ground vs. Raised Bed Gardens

The most water-efficient style for a vegetable garden is when the plants are planted directly in the ground. This allows more room for plant roots to grow so they can absorb more water and are shielded from the drying effects of temperature and wind.

Let's start at the beginning to determine which type of vegetable garden requires the least water. While raised bed gardens are easy to install and a popular option for homeowners, they use more water due to exposure to higher air temperatures, which causes the soil to dry out more quickly. Additionally, the soil is also more exposed to wind. Both wind and higher air temperatures increase the amount of water vegetables need. The taller the raised bed, the more water is needed as the soil will dry out more quickly; shorter sides—6 to 8 inches (15–20 cm) tall—is best. However, a taller raised bed may be your only option if you have poorly drained soil.

Different materials are used to construct raised beds, and wood works best as it doesn't heat up. Avoid using metal or concrete blocks unless you live in a region with mild summers because metal and concrete blocks get very hot and heat the soil, which causes water stress for vegetables and elevates their water requirements.

### *Vertical Gardening*

One of the newer garden trends is vertical gardening, which is a garden arranged vertically instead of laid out on the ground. This upright gardening style saves a lot of space; no bending or stooping is needed to tend to plants, and it is aesthetically pleasing. Vertical gardens can save water in climates with medium to high humidity levels, as water from the upper levels can filter down to lower levels. However, they aren't a great choice in drier climates because they need a lot of water to prevent the small root balls of the plants from drying out.

## Plant at the Right Time

It's important to follow the recommended vegetable planting calendar for your region. A common mistake people make is to plant too early or late in the growing season. Regarding water usage, planting warm-season vegetables too late for your climate stresses the vegetable plant as it has to work harder to grow under less-than-ideal temperatures and sunlight. The same is true if you plant cool-season fall crops too early when it is too hot outside. A vegetable planting calendar for your region is a great planning tool—ask your local nursery if they have one, join an online garden group, or look for online resources for your climate.

Actual climatic conditions also need to be considered when deciding when to plant vegetables. We are experiencing more unusual weather patterns with unseasonal temperature swings. If you are experiencing a heat wave in early fall, wait a week or two before planting cool-season vegetables. Use your weather forecast and a vegetable planting calendar to determine the best time to plant.

The author's husband and son incorporate compost into a new in-ground vegetable garden.

## Soil Preparation

Vegetables require loamy, nutrient-rich soil to grow their best. Good quality soil with lots of organic matter will hold onto water so it doesn't dry out quickly and will encourage deep rooting. Compost and high-quality topsoil are your best friends when creating good garden soil for your vegetables. The compost helps heavy clay soils to drain better and increases the water-holding capacity of sandy soils while also providing beneficial microorganisms. Be sure to use compost with a fine texture and that doesn't have woody pieces throughout, which can steal nutrients from vegetables as they break down over time. Prepare your new vegetable garden soil at least two weeks before planting.

### *In-Ground Vegetable Garden Soil Prep*

For new in-ground vegetable plots, select an area with good drainage (see chapter 4, page 60, on how to test for good drainage). Once you've ascertained you have good drainage, spread 3 to 4 inches (10 cm) of compost over the bare ground. Then add 1 inch (3 cm) of composted chicken, rabbit, or steer manure. Shredded leaves, mushroom compost, and worm castings also can be sprinkled on top of the compost. Mix the compost and other organic matter into the existing soil to a 6 to 8 inches (15–20 cm) depth. A rototiller is a helpful tool for creating a new vegetable garden area to incorporate the initial application of compost and other amendments, but a regular shovel works too. It's a good idea to test the garden soil and add other amendments for in-ground beds to help with nutrient deficiencies.

### *New Raised Bed Soil Prep*

If you have a raised bed, the soil mixture consists of 50 percent high-quality topsoil and 30 percent compost. Fill the remaining space with other

## WHY YOU SHOULD TILL ONLY ONCE

In the past, tilling garden soil every season to incorporate amendments such as compost and manure before planting was an accepted practice. However, studies have shown that this destroys soil structure, reduces water-holding capabilities, raises weed seeds to the surface, and increases erosion. Tilling should be done once when you add a new vegetable garden to mix in compost, manures, and other organic matter. Thereafter, at the beginning of each vegetable growing season, apply amendments on the top of the soil and lightly rake to mix them together without touching the soil underneath.

organic matter such as aged chicken, rabbit, or steer manure, shredded leaves, or worm castings—feel free to add just one of these materials or a combination to fill the remaining 20 percent. Mix all the materials with a shovel or rototiller. Over time, the soil amendments will make their way down to the soil below, improving it, which will benefit your vegetables too.

## What Type of Vegetables Are Best for a Waterwise Garden

Look through a seed catalog or a wall filled with seed packets at the nursery, and there are countless choices for each type of vegetable. It's easy to get overwhelmed by the sheer number of choices available. For every type of vegetable, there are many varieties that offer specific features desirable to the grower. These features vary and can include differences in size, flavor, color, cold or heat tolerance, and lower water requirements, to name a few. Often, characteristics will focus on certain types of climate; when possible, you should select a vegetable variety tailored to your climate.

### *Heirloom vs. Hybrid Vegetable Varieties*

There has been renewed interest in heirloom varieties of vegetables, which are old cultivars of vegetables grown long ago that continue to be grown. Heirloom vegetables are usually self- pollinating, meaning they self-fertilize but are occasionally cross-pollinated by insects. Modern vegetable varieties, also known as hybrids, are intentionally cross-pollinated by growers who use two different varieties of a particular vegetable plant to create a new one with the best features of each parent. Bees and other pollinators cross-pollinate plants naturally, but growers select which two plants to cross-pollinate with each other. For example, a tomato plant with delicious flavor is crossed with another tomato that produces more fruit to create a new plant with both features. It's important to note that hybrid vegetables are not GMOs (genetically modified organisms). Heirloom vegetables produce seeds that will produce a plant identical to the parent, while hybrids don't produce seeds with the same features.

Regarding water efficiency, hybrid vegetables have a slight advantage over heirloom varieties if bred to use less water. Heirloom varieties native to drier regions tend to use less water than those from wetter zones. Talk to growers at your local farmers market to see what varieties of vegetables perform best where you live.

A radish seedling emerges from soil amended with compost and aged steer manure.

***Growing Vegetables From Seed vs. Transplants***
Vegetables can be grown from seed and many from young seedlings (transplants). The benefit of growing from seed is you have more varieties to choose from than when growing from transplants, so you have a wider array of traits to select from. Additionally, seeds are much less expensive than purchasing transplants. However, they do require more frequent watering at the beginning until they begin to sprout. If you grow from seed, it will take longer for the plant to grow to maturity, so be sure to plant as soon as you can, either in the garden or get a head start by starting seed indoors. Transplants are young seedlings that come from the nursery and help to give you a head start in the vegetable garden. They are more expensive than purchasing seeds, and fewer varieties are available. It's advantageous to purchase vegetable transplants grown locally that are already adapted to current weather conditions. For example, a cucumber transplant grown in a cooler region and then shipped to a warmer climate may struggle initially, requiring more water. Don't be afraid to ask your nursery professional where the vegetables were grown.

## LOWER WATER USE VEGETABLES

Vegetables native to drier regions use less water than those from more temperate regions. Warm-season vegetables generally have higher water needs than cool-season vegetables. Here are vegetables that are water efficient:

- Asparagus
- Pole beans
- Snap beans
- Tepary beans
- Swiss chard
- Eggplant
- Melons
- Mustard greens
- Peppers
- Squash
- Tomatoes

Herbs such as basil, chives, rosemary, sage, and thyme are good lower-water additions to the vegetable garden.

A raised bed planted together in staggered rows to maximize the growing space and harvest.

## Intensive Vegetable Planting

Grow your vegetable plants close together, so as they grow, the soil will be shaded by the leaves, which helps to prevent the soil from drying out quickly. This intensive planting method has ancient roots and is still practiced worldwide to maximize the harvest from smaller plots while minimizing water waste and weeds. When you plan each season's vegetable garden, note how large each type of vegetable plant will grow and space it accordingly—the actual arrangement of where you grow your vegetables matters in terms of water usage. In a traditional single-row planting of vegetables, they are spaced to allow room for growth on two sides of each plant, with an empty space between the next row where water is wasted. With intensive planting, individual vegetable plants are planted in a staggered pattern so that each plant is equally distant from surrounding plants on all sides. This planting method allows the leaves of vegetable plants to touch and shade the soil, preventing rapid drying of the soil and conserving water.

### Intensive Vegetable Gardening

**Traditional Rows**

Planting traditional rows fits 10 plants in this bed.

**Square Grids**

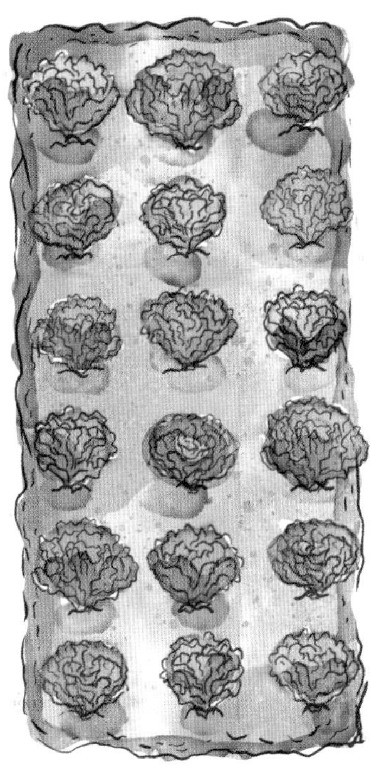

Square-foot gardening fits 18 plants spaced 1 foot (30.5 cm) apart.

**Intensive Gardening**

Using staggered rows for intensive gardening fits 23 plants.

## Mulch the Vegetable Garden to Maximize Water Efficiency

One of the most effective ways to maximize water efficiency is to apply mulch around your vegetables. Mulch keeps the roots cooler, so your vegetable plants need less water. It prevents the soil from drying quickly and keeps the water near the plants instead of running off. A bonus is that as the mulch breaks down, it adds nutrients and improves soil structure. The best types of mulch for vegetable gardens are straw or shredded leaves. After the seeds have sprouted and the soil has had time to warm up, water the entire surface of the vegetable garden and then apply a 2-inch (5 cm) layer of mulch on top of the soil. Do this every growing season. Don't pile mulch up against the base of plants. Keep it at least an inch away from the stems of each vegetable plant—a nice, even layer throughout the garden is best. A bonus of a well-mulched garden is fewer weeds, which can steal water from vegetables.

## Fertilizer

Vegetables require a high-nutrient soil to grow in, and supplemental fertilizing is often required—especially when a garden is new and the soil hasn't fully built up its nutrient-rich structure. Adding organic matter such as compost and aged manure contributes to soil fertility, but it

Straw mulch protects the soil around young beet plants from drying out.

takes time for nutrient levels to grow. Even then, applying fertilizer may be necessary throughout the growing season. Organic fertilizers are a more water-efficient choice than chemical fertilizers, which contain a lot of salts and increase a plant's water needs. You can select an organic fertilizer formulated for vegetables and follow the package's directions on when and how much to apply. You may want to do a soil test each year to see what nutrients your garden may need and which may be available in sufficient amounts.

## When and How Often to Water Vegetables

Water vegetable gardens in the early morning before the sun is fully up. Plants absorb water most efficiently in the morning before the sun and the day's heat set in. If you research how often to water your vegetable garden, you'll get general guidelines, but the problem is that they don't pertain to specific conditions and can cause water waste. Conditions that affect how often your vegetable garden will need water depend on several factors, many of which can vary weekly or seasonally. First, let's look at the seasons—the heat and sun of summer mean that the vegetable garden will dry out faster than other seasons. Temperature and humidity also factor in watering frequency—the hotter and drier it is, the more often you'll need to water. Local weather also affects how quickly soil and plants dry out. Windy conditions mean more water is needed, while rainy weather means you'll have to water less. The age and efficiency of your irrigation method can play a factor, as does the water pressure at your house. Raised bed vegetable gardens will dry out more quickly than those planted in the ground.

So, with all these variables affecting how quickly the vegetable garden dries out, how can we determine when to water? The answer is easy and involves a simple tool—your finger. We need to test the soil moisture in the garden to let us know when to water. To test, insert your finger into the soil to a depth of 2 inches (5 cm)—a small trowel works, too, if you want to keep your hands clean. If the soil feels moist, then no water is needed at that time. However, if the soil is mostly dry, you need water. You'll do the test more often (maybe daily) at the beginning, and over time, you will notice patterns and trends—specifically, how weather conditions and the seasons affect how quickly the soil dries out. Moving forward, you can implement those schedules based on your observations.

When you apply water, we want it to permeate 6 inches (15 cm) deep. Keep an eye on your plants. If you see signs of wilting, apply more water. When vegetable plants are very young, they need more water to help them grow a healthy root system. As they begin to mature, they need less water.

## Watering Methods for Vegetable Gardens

A consistent water supply to your vegetable garden is essential to successfully growing vegetables. Vegetables are best watered at the plant's base to limit water evaporation. If you use a hose or sprinkler to water vegetables, 20 to 50 percent of water is lost immediately to evaporation. Additionally, water on the leaves of vegetables can increase the incidence of damaging insects and fungal diseases. The good news is there are several excellent methods to maximize watering efficiency without undue waste.

### *Why Hand Watering Is Not the Best*

Many new vegetable gardeners devote little attention ahead of time to how much time and trouble it takes to hand water their garden. As discussed earlier, vegetables aren't low-water use and often rely on you for water needs. During the height of summer, watering once a day quickly becomes burdensome, and if you skip a day or two, it may mean the death of a favorite vegetable plant. Hand watering also leaves you chained to your garden, without the freedom to go on vacation, unless you have a reliable person to water your garden.

Watering with a hose produces uneven results, with water permeating deeply in one area rather than the other, depending on where you concentrate your watering efforts. Or, if you have large-leafed vegetables, water can sheen off the tops. If your water source is far from the vegetable garden, hauling your hose isn't fun.

If, however, watering by hand is the method you have to work with, use a rain wand to water vegetables and water them at their base instead of watering from above. To decrease your watering sessions, consider using ollas or creating a waffle garden that allows you to go for longer periods in between watering (see pages 102 and 150).

## Drip Irrigation

Drip irrigation is the most water-efficient method for irrigating vegetables and, when installed, is the easiest to manage. Water slowly drips from emitters at the base of individual vegetable plants and minimizes evaporation with water permeating down into the root zone without runoff. For vegetable gardens, you have several options to consider. The first is drip irrigation tape that has tiny slits or drip emitters embedded along the length of flat tubing, which slowly drips out water. The tape is laid in rows on the soil surface every 8 to 12 inches (20–30 cm). A similar method uses ¼-inch (6 mm) drip irrigation microtubing with small holes along the entire length. Another option is to use regular ½-inch (1 cm) black poly-tubing alongside each vegetable plant with a drip emitter inserted directly onto the tubing to administer water directly to the plant. Small metal stakes anchor both the tape (or tubing) in place.

### *Soaker Hoses*

Another irrigation method that works well in the vegetable garden is soaker hoses. A soaker hose is porous, so as water runs through it, it slowly leaks water into the soil at a slow rate, allowing water to permeate the soil at the base of the plants and helping to eliminate water lost to evaporation. Soaker hose kits are available at home improvement stores and require the same connection to a water source as drip irrigation.

Drip irrigation microtubing weaves throughout a raised vegetable garden.

## TEMPORARY WILTING

In the summer vegetable garden, you may notice some of your larger-leafed vegetables wilting in the afternoon. This type of wilting results from more water being lost from the leaf surface than the plant can replace. If your plant recovers in the evening without wilting, it doesn't harm them. Just be sure they are getting adequate moisture when you water.

Example of waffle gardening practiced throughout the American Southwest by indigenous peoples in the past and today.

## Water-Conserving Strategies from the Past for Today's Vegetable Garden

Methods used long ago to irrigate vegetables are still relevant today, long before drip irrigation or soaker hoses existed. Buried clay pots (ollas) and keyhole and waffle gardens are strategies that farmers long ago (and today) utilized in regions where water can be scarce. To help us minimize water loss in the vegetable garden, these irrigation methods are worth exploring and incorporating in your garden.

### *Waffle and Keyhole Gardens*

In the American Southwest, indigenous people created earthen berms to trap and channel water to direct it to specific crops. This technique is still practiced throughout drier regions of the world. Unlike berms in long rows, circular or square basins are created using 4-inch-tall (10 cm) edges (berms), made from existing soil, or dug into the ground, which captures water and keeps it for the plants. In regions affected by drought, waffle gardens are seeing a resurgence in popularity.

Keyhole gardens were used in drought-affected regions of Africa to make the most of the little water available and to overcome the poor soil conditions. These gardens consist of a raised, circular garden with a section cut out to allow for access to the garden's center. A hole (well) lined in chicken wire or wire mesh in the middle of the garden runs to the bottom. The well can serve as a compost bin through which you water your garden to keep the soil moist. See how to create a keyhole garden in chapter 11, page 178.

### *The Three Sisters Method*

The Iroquois and Cherokee peoples of North America utilized a companion planting of three different types of vegetables to help each other in the garden. The "sisters" consist of corn, beans, and squash. Beans use the tall stems of corn to climb up to reach

**top, left** Unglazed, earthenware pots, also known as ollas, buried in a vegetable garden that will provide water over several days.

**top, right** A waffle garden in Denver, Colorado, where individual cells concentrate water for vegetables. Creeping thyme grows on top of the berms, which enjoy slightly drier soil than the vegetables.

**left** The three-sister's corn, beans, and squash method on display at the Tohono Chul Gardens in Tucson, Arizona.

Self-watering raised beds have a reservoir that supplies water to the vegetables from underneath, reducing water loss from evaporation.

the sun. Beans take nitrogen from the air and put it into the soil to help add this vital nutrient, which in turn helps the corn and the squash. Large squash leaves shade the soil to complete the trio, cooling the roots and preventing the soil from drying out too quickly. Three sisters is a fun method to use in the summer vegetable garden, especially when you realize how these three plants benefit each other.

### *Wicking Vegetable Garden Beds*

A popular method of growing vegetables in dry climates is to use a self-watering raised vegetable garden bed. Water is added to a reservoir at the base, drawn up to the soil and vegetable plants above. Capillary action causes the water to move upward as the soil begins to dry out, providing consistent moisture for plants. Because the vegetables receive water from below, there is less water loss from evaporation versus surface watering. Additionally, less frequent watering is needed as the reservoir stores enough water for several days before needing to be refilled. For how to build a wicking garden bed, see chapter 11, page 176).

## Growing Fruit

In addition to vegetables, water-efficient fruit gardening is possible to do. Like vegetables, fruits aren't low-water plants but are watered less frequently than vegetables. The key to growing fruit is to ensure that you implement water-saving strategies.

### Fruit Tree Irrigation

For fruit trees, we want to water deeply and infrequently to foster deep root growth. In chapter 6, we discuss how to properly water trees to maximize a healthy root system, which also applies to fruit trees. Water should permeate to a depth of approximately 2½ to 3 feet (76–91 cm) for each water cycle. Drip irrigation works well for fruit trees and is the most water-efficient method. Basin flooding also

A sweet harvest from a peach tree that receives water via drip irrigation.

works but is less water efficient than drip irrigation. This technique creates a basin around the tree's root zone and edged with a 4- to 6-inch-tall (10–15 cm) berm of soil. The basin is filled with water to the top that soaks slowly into the soil. You may need to fill the basin twice to reach the recommended watering depth.

## Mulching and Fertilizing Fruit Trees

Apply a 2-inch (5 cm) layer of organic mulch around the tree's root zone that extends out to the outer branches every spring. Use compost, shredded leaves, straw, pine needles, or wood chips for mulch. Keep the mulch several inches away from the base of the trunk to prevent conditions favorable for fungal disease. As the mulch breaks down, it will enrich the soil, building up its structure. The fertilizer needs of fruit trees differ according to the type of tree and when it produces fruit. In warmer climates, fruit production occurs earlier in the year than in cooler regions. Consult a garden expert in your area to learn when to fertilize, which is timed based on when fruit appears. Use organic fertilizer tailored to the needs of your tree. Avoid using chemical fertilizers, which will increase the water usage of your tree.

## Prune Off Unproductive Growth

To reduce the water needs of your fruit trees, prune away old branches that may no longer produce any fruit. The branches that remain should be relatively young and produce fruit. Old, unproductive branches increase the tree's water needs, and if they aren't contributing to the fruit harvest, you can remove them unless they are producing needed shade for lower branches. Pruning is limited to certain times of the year. Don't prune in summer, which increases a tree's water usage. Most fruit trees, like apple and stone fruit trees, are pruned in late winter, while citrus trees are pruned in spring once the threat of frost has passed.

## Thin Your Fruit

Trees laden with a lot of fruit use a lot of water. Thinning is a common practice for fruit trees such as apple, apricot, nectarine, peach, pear, and plum trees. This involves the removal of excess fruit to decrease the nutrient and water load of the tree. As a result, the remaining fruit grows larger and is of better quality as it isn't competing with larger numbers of fruit for the tree's limited resources. Thinning should happen when fruit begins to form and is still relatively small. Pluck off excess fruit so there is at least a 6-inch (15 cm) spacing between the remaining fruit. Thinning reduces the water needs of fruit trees by up to 20 percent to 30 percent.

Unripe peach fruit that needs to be removed so that only one fruit remains.

CHAPTER 11

# Projects

Flowering groundcovers such as purple and white lantanas (*Lantana montevidensis*), angelita daisy (*Tetraneuris acaulis*), with coral fountain (*Russelia equisetiformis*) blooming against the wall, add colorful interest where a thirsty lawn used to be.

# Lawn Removal

Removing your lawn is the best way to save water in your garden. As discussed in chapter 8, a grass lawn needs frequent watering and maintenance to keep it healthy and weed-free. If your lawn serves no purpose other than decorative, you can create an even more beautiful and diverse outdoor space using plants such as groundcovers to create a similar look.

Taking out your lawn is a laborious process, yet the rewards are immediate, and there will be less maintenance in the future and a substantial reduction in water use—making it worth it.

The plants you use to replace the lawn will add beauty and create a habitat for bees, birds, and butterflies while requiring a fraction of the water and work to keep them looking their best.

## Lawn Removal Options

When removing your grass lawn, different approaches depend on variables such as ease of use, effectiveness, type of grass removed, and your climate. The most popular removal methods are digging up grass, solarizing, and sheet composting. Let's explore each option so you can select the best option for your needs.

### Digging Up the Lawn

The title of this option is self-explanatory—this is the fastest method for lawn removal, and it involves digging up your lawn using a shovel, sod cutter, or rototiller. Using a shovel is the most labor intensive as it involves a lot of physical labor and takes the most time. A flat shovel is preferred so you don't create an uneven surface when digging out the grass.

For a less labor-intensive job, a sod cutter is a good option. This lawn removal tool is usually available for rent from your local home improvement store, or you can hire a landscape company to come and do this for you. The sod cutter is a machine that cuts out sections of grass, making removal more efficient and easier. Once the sod cutter has cut out the grass sections, you can discard the lawn in neat sections. Follow up with

A high maintenance, thirsty lawn before removal.

Lawn removal in process. Irrigation was stopped a month prior to removal, and a sod cutter has cut out the grass in sections.

A sod cutter cuts out rows of turf for easier removal.

a flat shovel to remove any remaining grass near borders. I recommend passing over each lawn area twice with the sod cutter to help entirely remove the roots to decrease the chance of grass growing back from the roots.

Another standard option for digging out a lawn is to use a rototiller. A rototiller has circular blades that work to turn the soil over and is often used to till compacted soil. When used for grass removal, the rototiller breaks up the grass into clumps mixed with soil, and you will have to sort through them to separate the grass from the soil. However, while a rototiller does work to remove grass, it can bring up weed seeds near the surface and also can leave grass roots behind, which may grow back. For vigorous grass species such as Bermuda grass, a sod cutter or shovel is a better option as this vigorous grass grows back from the smallest root left behind. Additionally, rototilling doesn't leave a smooth surface afterward, as when you use a flat shovel or sod cutter.

Whichever method you use to dig up the lawn, make sure to dig down 4 to 6 inches (10–15 cm) to remove as many of the roots of the grass as possible to prevent it from growing back. After removing your lawn, you need to look for spots of grass growing back. If any roots remain in the soil, grass will likely re-emerge. So, be on the lookout and remove any grass that comes up while it is still small.

### *Herbicides*

Herbicides (weed killers) are often used to pretreat a lawn before digging it up. However, *you can remove a lawn without using them*. Herbicides are chemicals used to remove or control unwanted plants and are typically applied two weeks before removing the grass to allow them time to work to kill the grass, including the roots. The key to success without using herbicides is to dig out the lawn deeply to a depth of at least 6 inches (15 cm) to remove all grass roots, keep an eye on any areas where blades of grass try to re-emerge, and pull them out right away.

## Solarization

This method involves more patience and uses the sun's power to kill your grass with heat and occurs in the summertime. Solarization involves covering your lawn with a clear plastic sheet (tarp) for several weeks in the summer. It's essential to do it when the weather is at its hottest and in an area that receives full sun for at least six hours a day to be most effective. The heat from the sun is intensified and trapped under the plastic, which heats the soil, killing the roots of the grass. This process takes patience, as killing the grass can take at least six to eight weeks. Solarizing soil in regions with cool and cloudy conditions isn't effective, and other lawn removal methods are best.

You will need enough clear plastic (1.5 to 2 mil thick) to cover the lawn with an extra 8 inches (20 cm) for overlap. The thinner the plastic, the more the sun can penetrate to heat up the soil. Additionally, you'll need bricks, rocks, or large staples to help anchor the plastic, which helps to trap the heat to prevent it from escaping. Once you have your

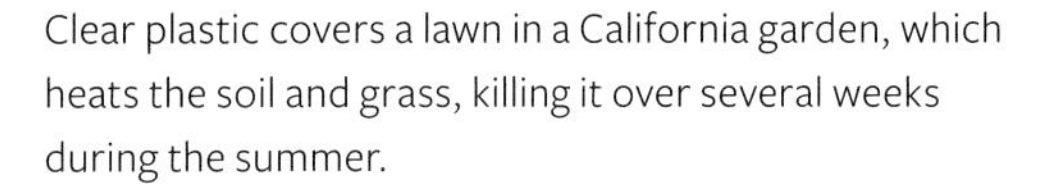

Clear plastic covers a lawn in a California garden, which heats the soil and grass, killing it over several weeks during the summer.

An area of grass sheet mulched, with a layer of cardboard and a thick layer of straw.

materials, mow your grass as short as possible. Now, it's time to water deeply to a depth of 12 to 18 inches (30–46 cm), which provides the moisture needed to maximize the heat under the plastic. Turn off your irrigation system. Cover the lawn with plastic sheets, overlapping each section by 8 inches (20 cm), and anchor the sheets using bricks or rocks. Avoid walking on the plastic, which could create small tears.

The hard part is done—now all you have to do is wait six to eight weeks. Many people recommend keeping the grass covered for only four to five weeks, but this may not be long enough to kill all the grass (longer is better). You will see the grass turning yellow and then brown. Wait at least two weeks after it turns brown before removing the plastic. If you remove the plastic too soon, the grass may grow back. After removing the plastic, you can clear out the dead grass with a shovel or sod cutter (no rototiller). However, it is okay to let the dead grass remain, add new plants, and cover it with mulch—the grass will break down naturally.

Solarization may not kill all vigorous grasses such as Bermuda grass since their roots grow very deep, and the heat may not reach them. Keep an eye out for any emerging grass and remove it right away to avoid it spreading.

## Sheet Mulching

Sheet mulching (sheet composting) is the least labor-intensive method for killing grass lawns. It entails putting a sheet of cardboard, brown builders' paper, or newspaper over the lawn and then adding a thick layer of organic mulch over it. Over time, the grass is killed due to a lack of sunlight, and the barrier material breaks down. However, it is not the best option in all situations, especially in dry climates. It's also the least water-efficient removal method.

This method is a popular way to remove a lawn because the materials needed are easy to procure and it requires little digging. The soil is enriched with organic matter from the cardboard as the grass and sheet materials break down over several months. However, this process depends on relatively even moisture, which can be problematic in dry climate regions because the sheet material won't break down well due to lack of rainfall unless watered by hand frequently. However, sheet

Before and after photos of a front yard in Mesa, Arizona, that has had the lawn removed and converted to an attractive xeriscape landscape.

mulching is a good option for removing grass for vegetable gardens (even in dry climates) because they are kept evenly moist from the irrigation from watering the vegetables.

If you have vigorous weeds or grass species such as Bermuda grass, sheet mulching is less effective than other methods for lawn removal as they can survive a very long time (more than a year) without sunlight and frequently re-emerge. I recommend consulting with local garden experts for the best option for your region.

Sheet mulching is best done in the wetter months, such as fall, winter, or spring. To begin, create a 4-inch (10 cm) shallow trench around your lawn area to allow a place for the mulch to go without running over the edge. Mow your lawn very short and then water it to at least 12 inches (30 cm) deep. Spread a 1-inch (3 cm) layer of compost over the wetted grass area. The best cover materials are newspaper or brown builders' paper from your home improvement store. Corrugated cardboard, especially heavy-duty, isn't always the best option; it is often coated inside, making it hard for it to break down, which inhibits water from permeating and the soil from breathing. Working in sections, lay a single layer of builders' paper or seven layers of newspaper down, overlapping as you go. Lightly spray with water as you go and cover with a 4- to 5-inch (10–13 cm) layer of wood chip mulch or compost. Keep working in sections until you are finished. If any plants are present, keep the cardboard/newspaper 12 inches (30 cm) away from the base of the plants. Keep the covered area evenly moist by sprinkling with the hose every few days, in the absence of rain, for at least three months.

## Before Lawn Removal and After

The benefits of lawn removal are immediate, and you'll enjoy the knowledge that you are saving water while decreasing the maintenance of your garden. Whichever method you use to remove your grass, you must be vigilant in the summer months, for the next year or two, for spots where grass may try to re-emerge. Pull grass out as soon as you see it. Eventually, the roots will starve, and the grass will cease to appear.

# Portable DIY Drip Irrigation Bucket

Plants need water to grow, and if you need to provide supplemental water to your plants, how can you water them most efficiently without an irrigation system? Often, this means watering by hand using a hose or watering can. But before you grab your watering can or attach your hose to a sprinkler, let's look at an option that uses less water. As discussed in chapter 6, drip irrigation is the most water-efficient method for watering plants. However, suppose installing a drip irrigation system isn't in your immediate future, or you want to provide extra water to a few plants. In that case, you can create your portable version of drip irrigation. Drip irrigation via these portable vessels is easy to make and allows you to provide needed irrigation for plants efficiently.

## DIY Drip Irrigation Supplies

The items needed to create a portable drip irrigation system are inexpensive, and you may already have them around your house. You will need the following:

- 5-gallon (19 L) (or smaller) plastic bucket with a removable lid
- 16-inch (41 cm) section of ¼-inch (6 mm) poly tubing
- Adjustable drip emitter (that can be turned off)
- Drill with 9⁄64-inch (4 mm) drill bit (or a similar size)
- Waterproof silicone caulk (optional)
- Needle-nose pliers (optional)

## How to Make A Portable Drip Irrigation Vessel with Plastic Bucket

Use a clean plastic bucket. If it has been used to store substances such as paint or other chemicals that may harm plants, it's better to purchase a new bucket with a lid, which is inexpensive and available at a home improvement store. To begin, mark the outside of the bucket with a permanent marker to indicate each gallon level to allow you to fill the bucket with the exact amount of water you want to apply to a particular plant. Next, drill a hole 1 inch (3 cm) up from the bottom of the bucket using a 9⁄64-inch (4 mm) drill bit or one similar in width to

Items needed to create a 5-gallon (19 L) bucket portable drip irrigation vessel.

A 5-gallon (19 L) plastic bucket, with ¼-inch (6 mm) poly tubing that ends with an adjustable drip emitter that can be turned off and on as needed.

A completed portable drip irrigation bucket providing extra water to an aloe plant in the author's garden.

the ¼-inch (6 mm) poly tubing. If the hole needs to be bigger, move the drill in a circular motion to increase the width of the hole to fit the drip tubing.

Once the hole is made, cut one end of the ¼-inch (6 mm) poly tubing at a 45-degree angle and insert the angled end into the hole so that at least 12 inches (30 mm) remain sticking outside from the bucket—you may need needle-nose pliers to help pull the tubing through the hole. Next, insert the adjustable drip emitter onto the outside end of the poly tubing—that is where the water will come out of to water the plants.

Place the bucket next to the plant you wish to irrigate and fill the bucket with the appropriate amount of water. Turn on the adjustable drip emitter to release water at the desired rate. If you see water leaking from around the hole in the bucket, empty it and let it dry completely. Then, apply waterproof silicone caulk around the ¼-inch (6 mm) poly tubing and allow it to dry twenty-four hours before use.

Once you have filled the bucket with water to irrigate a plant, loosely cover it with the plastic lid to prevent evaporation. Don't fit it tightly over the top, or that will halt the water flow. If you have plants located closely together, you can add another hole with tubing on the other side of the bucket to water two plants at once. This drip irrigation method is portable; you can move the bucket where needed and store it away when not in use. You can create multiple buckets to allow you to water a large number of plants at once.

## A 1-Gallon (3.8 L) DIY Irrigation Vessel

If your irrigation needs are smaller and a large bucket isn't needed, then a 1-gallon (3.8 L) plastic container may be more suitable. They are easier to move around and work well for smaller plants that typically need a single gallon or less of water for each watering cycle. You can use recycled gallon containers such as plastic milk or water jugs that are thoroughly rinsed out. You only need a few supplies for this easy project:

A 1-gallon (3.8 L) plastic container with holes on the bottom to drip water to the root zone of a new plant.

***Materials***

- 1-gallon (3.8 L) thin-walled plastic container with lid
- Small nail
- Needle-nose pliers
- Lighter or other flame

***To assemble***

1. Hold the tip of the nail with pliers and heat the sharp point of the nail with a flame for several seconds.
2. Take the nail and pierce the bottom of the plastic container.
3. Make three to four tiny holes—you may need to reheat the tip of the nail after each hole.

To use, place the plastic container next to the plant, fill it with water, and loosely place the lid on top—don't tighten the lid, or water won't flow out. Or, you can fill the container at your water source and carry it to the plant you wish to water. You can use these smaller drip irrigation vessels for watering container plants or use several for plants that have higher water needs.

# Make Your Own Olla for a Self-Watering Container Garden

Container gardening is very popular due to its versatility in a large garden or a tiny space such as a balcony or front porch. However, providing regular water to our container gardens can become cumbersome—especially if a water source isn't nearby. Hauling a hose or watering can isn't fun. But what if you could decrease how often you water your pots? Imagine the freedom of not having to water frequently and being able to go away for several days without worrying about your plants drying up.

Self-watering containers are a way you can do that, and you can purchase one that does this, but you can also turn a regular pot into a self-watering container. You can make an olla to self-water your potted garden or use it for in-ground plants, such as higher water plants or in your vegetable garden. Ollas are unglazed terracotta vessels that you bury in the soil, slowly leaching water to plant roots. They are a very efficient way to water container plants since they receive water from below, so less water evaporates versus overhead watering from a hose or watering can. You can purchase ollas, which cost anywhere from $30 USD to over $50 USD each, or make your own. The materials are easy to find, they are easy to make, and they are inexpensive.

## How to Make an Olla

### *Materials*

- Two identical, unglazed terracotta pots; plus one saucer
- A quarter or another coin large enough to cover the drainage hole
- Waterproof silicone caulk
- Waterproof glue

Water is added through the top pot's drainage hole, filling the container that slowly migrates into the surrounding soil over several days. The size of your DIY olla will vary, depending on the size of your container. In general, an olla will water plants two times its diameter. So, for example, an 8-inch-wide (20 cm) olla (at the top) will provide water 16 inches (41 cm) from each side. It's essential to allow enough room in the pot to add plants around the olla. The larger the pot, the bigger your olla should be, which will hold onto more water, and you will need to refill it less often.

Supplies needed to create an olla.

A completed olla ready to be buried in a container, in the ground, or in a vegetable garden.

A homemade olla is filled with water, which decreases watering frequency for newly planted hummingbird bush (*Hamelia patens*) and purple lantana (*Lantana montevidensis*).

*Steps*

1. Use the waterproof glue to adhere a coin that fully covers the drainage hole of one of the pots—leave the other hole open. Allow the glued coin to dry for twenty-four hours, and then fill the pot with water to ensure no water leaks from the drainage hole.

2. Next, run a line of clear silicone caulk around the top rim of one of the pots and place the rim of the second pot so they adhere together. Place a brick or other heavy object on the top while it dries to ensure a watertight seal. Allow it to dry until the next day.

3. Before adding your olla to the container, fill it with water through the open drainage hole to test that the seal is watertight. If not, apply more silicone caulk around the seal that joins the pots together.

4. Once your olla is fully dry and watertight, bury your completed olla with the sealed drainage hole end pointing downward in your pot. Leave the top inch or two of the olla above the soil line, where it is easily accessed to fill with water.

5. Add plants around the olla, ensuring they are within reach of where water will leach out from the olla.

6. Fill the olla with water from your hose or a watering can with a narrow nozzle, and then place the saucer lid over the opening to reduce evaporation and keep mosquitos out.

## When to Add Water to Your Olla

Test the soil moisture by inserting your finger 2 inches (5 cm) into the soil. If the soil feels barely moist, it's time to refill your olla. Remove the lid, fill the vessel with water, and replace the lid. Water will slowly leach into the surrounding soil, watering plants around it, reducing watering frequency. Variables such as weather, the size of the olla, and container size will affect how often you need to fill the olla reservoir. The olla will empty more quickly as the weather warms, so keep an eye on it. In hot, dry climates, you will need to refill the olla more often than in humid or temperate regions. Ollas can crack if you live in a cold winter climate that experiences freeze/thaw cycles. To avoid this, unbury your ollas and bring them into an area protected from freezing temperatures in autumn. See chapter 6, page 103, for more olla care instructions.

## DIY Self-Watering Container with Recycled Plastic Containers

Another DIY option for self-watering containers is repurposing plastic containers, such as water jugs or large plastic bottles, to create a buried watering vessel (similar to the olla). Make seven to eight small holes around the bottom half of the container using a small nail (heat the tip of the nail with a flame). Bury your plastic container in the center of the container, leaving the top inch above the soil with its lid. Fill with water and lightly place the lid on top—don't tighten the lid, or water won't flow.

## OLLAS ARE NOT BEST FOR ALL PLANTS

It's important to note that ollas are best utilized for plants that require regular water. Low-water plants, such as cacti and other succulents, or extremely drought-tolerant shrubs don't do well with ollas in container gardens since they do best with periods of allowing the soil to dry out between watering.

# The No-Water Garden

A Phoenix, Arizona, garden filled with native brittlebush (*Encelia farinosa*), creosote bush (*Larrea tridentata*), and foothill palo verde tree (*Parkinsonia microphylla*) doesn't require supplemental water other than rainfall. The brightly colored bougainvillea bush (*Bougainvillea* spp.) is highly drought-tolerant once established but appreciates the extra rain that falls from the home's eaves.

Imagine an attractive outdoor space filled with plants that receive all their water needs from rainfall. In many regions with plentiful rainfall, this is the norm where supplemental watering is focused on container gardens and during below-average rainfall. However, in climates where drier conditions reign and rainfall is less abundant, it can be more of a challenge to meet plants' water needs. This is particularly true when the plants in the landscape require more water than rainfall will supply. Is it possible to have a garden that adds beauty that can exist on the water brought to it by the rain? The answer is a resounding yes!

## How to Create a Rainfall-Only Garden

Planning and research are involved in creating a garden that can exist on average rainfall amounts to ensure that you make the proper decisions that will lead to success. These decisions involve knowing your average rainfall amounts and times of year rain falls. Once you know the amount of moisture that falls from the sky, you can move toward choosing plants that can exist on that amount of rainfall. Lastly, incorporating water-harvesting strategies can maximize the amount of usable water from rain, extending its use instead of letting it drain away.

### How Much Rain Do You Get Annually?

To determine how much rain falls where you live, make use of the online sources in the Resources section. Find out if your area is experiencing a drought. Rainfall amounts are lessening in many

Wildflowers such as California poppies (*Eschscholzia californica*) are an excellent choice for the water-free garden.

A low-water garden ready to be planted around a swale, which will receive much of its water from rainfall.

regions, so finding out how much rain has fallen in the past five years is helpful. This information is easily accessed online. Once you know how much rain you receive each year, we also need to find out what times of year the rain comes. Most regions have wetter and drier seasons. This is important to know, particularly if you have longer dry seasons.

## Plant Choices for the Water-Free Garden

All plants need water. But, if your goal is to create a landscape that needs little to no supplemental irrigation, we must select plants adapted for the amount and timing of the rain we receive where we live. The easiest and best option is to look first at using plants native to your region, adapted to average rainfall amounts and when it falls. If you have natural areas where you live, observe what is growing. Take a hike through the surrounding terrain and take pictures of the plants you observe. You can do online plant research or talk to your landscape or nursery professional to identify the plants you see. Botanic gardens are helpful resources as many have native plant areas for you to visit and observe the plants, which usually have plant signage to identify what they are.

Look for "nativars," which are cultivars of natives that plant breeders develop to increase optimal characteristics such as appearance and size while increasing temperature tolerance. In some cases, nativars use less water than the regular native plant. Finally, look to plants from other regions adapted to the climatic conditions where you live—where rainfall and its timing are similar to your region. This can expand your plant palette, although there is more variability in the success when you bring in outside plants from other regions.

### *Drought's Impact on No-Water Gardens*

During periods of drought, where rainfall amounts are lower than usual, you will likely need to provide supplemental irrigation for your plants. Look for signs of drought stress such as browning foliage, wilting, wrinkling, unseasonal leaf drop, or lack of flowering. Portable drip irrigation vessels (see page 162 in this chapter) can be helpful in times of drought where infrequent watering is needed.

## Maximize Plant Choices with Rain Harvesting

When we base our no-water garden on rainfall amounts alone, it assumes that the soil absorbs

Connected swales and a 3-inch (8 cm) layer of mulch will help these native and adapted plants survive with minimal supplemental irrigation once established. Non-irrigated gardens aren't all or nothing—they can exist near other areas that do require regular irrigation.

some rain while the rest drains away. However, incorporating active and passive rain harvesting methods increases the amount of water available for plants. In that case, it expands your plant palette to include those that need slightly more water than your region receives. Rain harvesting makes the most of rain, slows it down so it doesn't drain away quickly, and allows plants and the soil more time to use it. See chapter 5 for a variety of ways to harvest rain.

## Planting the No-Water Garden

Planting is stressful for new plants as they get used to a new location, sun exposure, and soil while growing. New plants will need supplemental water to help them become established—this can take a year or more. The success of the no-water garden is dependent on plants with a healthy root system, which enables the plant to absorb water from the soil when it does rain. Young plants have very small root systems and must be watered more often to help them grow enough roots to support the plant later on rain alone. Create holes at least twice the size of the root ball to make it easier for roots to grow outward. A shallow basin around each plant, extending to the mature width of the plant, can help concentrate water around the root zone when rain falls. To minimize how much water plants need when planted, the time of year you plant matters. While spring is the most popular time for people to add new plants, it's close to summer, which means young plants will need more frequent irrigation since their root systems are still small. Plant in fall when possible, which allows plants three seasons to grow roots before the heat and higher water stress of summer arrives. In mild winter regions, winter can also be an excellent time to plant.

### *Plant Care for Non-Irrigated Gardens*

Add mulch around plants, especially new plants, to prevent water from drying out too quickly. Less maintenance is more beneficial for gardens that exist without supplemental irrigation. When leaves and flowers fall, allow them to remain at the base of the plant to act as a mulch. This natural mulch layer is how native plants survive on rain alone, with their fallen leaves/flowers serving as a natural mulch layer. Be ready to provide supplemental water during drought or record-breaking summer temperatures if plants show signs of drought stress. Avoid frequent pruning, which increases a plant's water needs as it regrows foliage.

# How to Make a Rain Garden

A rain garden in Michigan lined with river rock that filters out pollutants and allows water to permeate the soil instead of running off into the street.

A rain garden is a popular method of harvesting rainwater to direct runoff around your property into a shallow swale with deep-rooted plants. The principle of this type of garden is to allow rain to soak into the soil to water plants and reduce water runoff and flooding. The more water plants receive from rain, the less often they need supplemental irrigation. Additionally, rain gardens help filter out water pollutants and attract wildlife, including pollinators such as birds and butterflies. Rain gardens aren't just for homeowners. Business and municipalities are incorporating this curb appeal friendly, water-harvesting technique.

## Making a Rain Garden

Rain gardens are a type of passive water harvesting that is perhaps the easiest to implement and makes an attractive addition to the waterwise garden. You can have one rain garden or several, depending on the typography of your landscape and the amount of runoff. The rain garden averages 6 inches (15 cm) in depth but can be slightly shallower or a little deeper. While this type of garden is pretty straightforward to create, you need to check if a permit is required to add one in your municipality. Then, we need to ensure it is in the right location and that the soil will absorb rainwater quickly. So, let's get started!

### Location Matters

The first step to creating a rain garden is to observe where runoff occurs when it rains. Note where the water flows through your landscape and any low areas where it may tend to pool. These are typically great locations for a rain garden. As with all passive water-harvesting methods, be sure to locate the garden at least 10 feet (3 m) away from buildings to prevent flooding. Don't locate the rain garden too close to existing trees or large shrubs, as it can damage roots. It's also important to ensure that there are no underground utility lines in the path

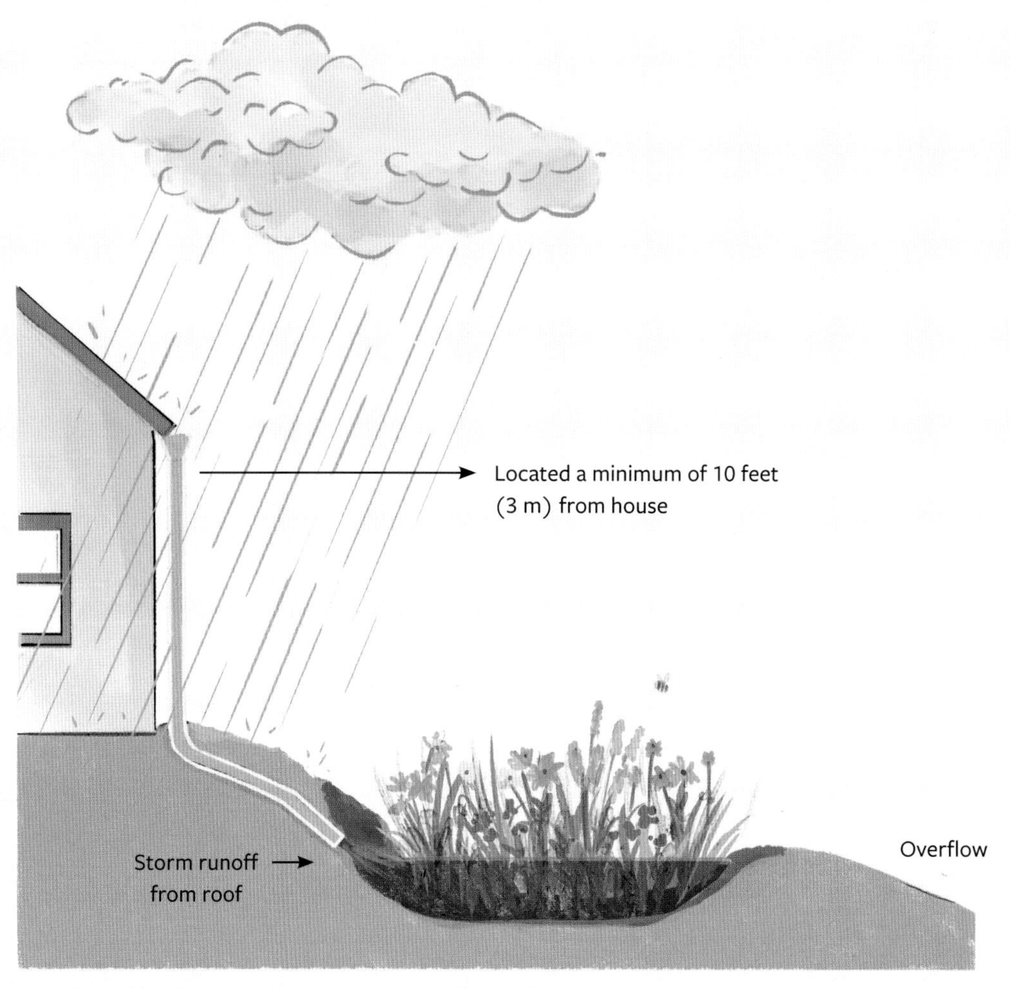


Water is used by plants and the excess moisture infiltrates down to groundwater level.

of your future rain garden. Also, locate the garden at least 25 feet (7.6 m) from basements, septic tanks, and wells.

Before you begin creating your rain garden, we must ensure adequate drainage in your selected location. If soil is slow-draining or tends to stay wet, that isn't a good spot for a rain garden. Water within a rain garden must soak into the soil within twenty-four hours to avoid mosquito breeding. To determine adequate drainage, dig a hole in your chosen spot about 1 foot (30 cm) in depth and fill it with water. You can move forward if water has soaked into the soil within a day. However, if there is still water in the hole, it's a good idea to find another location, or you can amend the soil with compost and perform the test again to see if you have sufficiently improved drainage.

## Rain Gardens—Small or Large?

One of the many desirable characteristics of rain gardens is that they come in all sizes—big, medium, and small. The size of your rain garden corresponds to the amount of water it can hold before overflowing. The construction of a rain garden allows it to overflow once it has reached capacity. So, if your runoff isn't huge, then a 100 square-foot (9 sq. m) rain garden may be sufficient. Alternatively, a larger size may be more helpful if you experience a lot of rain draining away from your home or construct several connected rain gardens. Residential rain gardens can be as big as 400 square feet (37 square meters).

To determine the size of your rain garden, we need to identify the amount of rainwater runoff that ends up in the rain garden. To do this, calculate the rain that comes from the downspout of your rain gutter, which can include runoff from the entire roof or a partial section of the roof. Use the following calculation to ascertain the maximum size of a 6-inch-deep (15 cm) rain garden:

- Multiply the length and width of the roof section to get the square footage of the catchment area. Example: roof section of 20 ft × 30 ft (6.1 m × 9.1 m) = 600 square feet (55.7 square meters).
- The rain garden is typically 20 to 30 percent the area of the catchment area.
- Example: 20 ft × 30 ft (6.1 ft × 9.1 m) (rooftop area/impervious surfaces) = 600 square feet (55.7 square meters) (your impervious rooftop surface area) × 0.20 (percentage of the catchment area) = 120-square-foot (11.1-square-meter) rain garden

In the equation above, you can include other impervious surfaces around your home besides the roof, such as a driveway, patio, and sidewalk.

You don't have to create the maximum size rain garden—adding a smaller rain garden is okay, or you can add several connected rain gardens that are sloped so that the top one overflows into the lower one, and so on.

## Building Your Rain Garden

Once we know our location and the size of our rain garden, it's time to begin digging. Our goal is to excavate the soil to the maximum depth at the center, with gently sloping sides. The depth of a rain garden can range from 4 inches to 8 inches (10–20 cm), with an additional 4 inches (10 cm) to allow for compost and mulch. For a large area, a small backhoe will make the job easier, but a regular shovel is sufficient for smaller gardens. Amend the soil with 2 to 3 inches (5–8 cm) of compost to improve drainage, tilling it into the soil to a depth of at least 6 inches (15 cm). Remember that we want to make it easy for water to flow into the top edge of the rain garden, so don't build up a berm in that area. On the lower side of the garden—the one that points away from your home or impervious surfaces—we want to mound the excavated soil to create a berm to allow water to collect in the rain garden. This lower edge must sit below the height of the upper edge of the garden so that water can flow out once it fills with water. Create a channel from the water source, such as your downspout, to guide the water to the rain garden (see page 74). Use larger rocks, such as 4- to 6-inch (10–15 cm) river rock, at the garden's edge where water enters and flows out to prevent soil erosion.

## Planting the Rain Garden

The types of plants we add to the rain garden are important. We want to use plants with deep root systems, which will help open the soil and help water filtrate downward. Grasses (not lawn grasses), perennials, and small shrubs are best suited for this type of garden. Plants should be planted toward the center of the garden, where the depth is lowest. Young plants will need supplemental water until they become established for at least the first year. In dry climate regions, you may opt to add drip irrigation to the garden to turn on during dry times of the year and off throughout the rainy seasons.

Once plants are in, it's time to add mulch. The type of mulch you select matters—light organic mulches will wash away, so it's best to use a heavier material such as shredded bark. Another bonus is that shredded bark doesn't decompose quickly, so you don't need to replace it often. Rock mulch is another option for the rain garden. Utilize at least 3 inches (8 cm) of rock mulch—river rock is a great option that adds a decorative element.

A large rain garden in Minneapolis, Minnesota, with a rock-lined channel along the back slope to prevent overflowing.

## Rain Garden Maintenance

While rain gardens don't require a lot of upkeep, some tasks are necessary to enable them to work efficiently. Prune plants as needed. Provide supplemental water for plants during periods of drought or dry times of year. Replenish organic mulch as needed. For rain gardens with rock mulch, soil and debris can collect over several years, partially burying the stone. To rectify this, lift out the stones, clean them off with a strong jet of water from the hose, and place them back in the garden. If the sides of the rain garden have eroded over time, build them back up to their original height with soil.

# Water-Saving Vegetable Garden Projects

## Wicking Beds

Wicking beds are becoming more popular in dry climate regions as they use much less water than gardens with traditionally grown vegetables in the ground or in raised beds. It's helpful to consider them self-watering containers expanded to the vegetable garden because wicking beds are irrigated from underneath, not from the top. Watering from below versus above reduces evaporation, so vegetables use less water—about 50 percent less than a raised bed watered from above. A wicking bed is simply a raised vegetable garden with a refillable reservoir underneath the water that "wicks" up into the soil above via capillary action. It works much like a self-watering container, just on a larger scale, and decreases the frequency of watering vegetables.

The author's sister and niece harvesting tomatoes in their California desert garden.

### Benefits of a Wicking Bed

Wicking beds aren't just for people who live in arid regions. In addition to using less water to grow vegetables, the underground irrigation within a wicking bed reduces the risk of fungal diseases or sunburned leaves on sunny summer days that come from watering by a hose or above-ground irrigation. The soil surface of a wicking bed is drier than in a traditional garden bed, so weeds are reduced. The time between filling the reservoir varies depending on the weather but can go as long as ten days or more. In warm and drier parts of the year, refilling the reservoir will occur more frequently, up to once or twice a week—but compared to daily watering, that is much less! Wicking beds provide an even soil moisture level, which causes less stress for vegetable plants and can lead to improved yields. As with other raised beds, wicking beds allow you to grow vegetables where existing soil texture or nutrients are poor. Wicking beds require less fertilizer than other beds since nutrients aren't flushed away. Finally, the pond liner prevents roots from other plants from infiltrating the bed.

#### *Disadvantages of Wicking Beds*

While there are many great reasons to consider adding a wicking bed to your garden, there are some drawbacks. The first is that creating a wicking bed is more expensive than building a regular raised bed, requiring several different components. Consequently, their construction is more involved and takes more time and effort. You can purchase a wicking bed kit, which decreases the work to create one, but at a higher cost than doing it yourself. Young seedlings or transplants will likely need supplemental irrigation from the top until their roots grow deep enough to reach the moister soil. Since they have a constant moisture supply, vegetables grown in wicking beds have smaller root systems than those grown in typical raised beds. However, in hot, dry climates, the smaller root ball can cause leafy green vegetables to temporarily wilt during the day's heat. Vegetables such as garlic, which do best with some periods of dry soil, may struggle to grow in a wicking bed. Additionally, perennial fruits and herbs aren't recommended in wicking beds as they can be sensitive to higher salt levels in the soil, whereas annual vegetables are fine.

## Building a Wicking Raised Bed

The benefits of creating your wicking bed and the rewards are many. Let's follow the steps you need to build your self-watering vegetable garden. First, we need to consider the location of our bed. Wicking beds share the same requirements as regular raised beds regarding sun exposure. Second, we must build our beds on a level surface to allow water to permeate the bed evenly without puddling. Determine what length you want your wicking bed to be, which can range from 3 to 10 feet (91 cm–3 m), with the width at 3 to 4 feet (91 cm–1 m). We are ready to gather our supplies once we know where our bed will go and its size.

### *Materials*

- Wood planks for the bed—2- × 12-inch (5 × 30 cm) pine boards or other untreated wood work well
- Plastic pond liner that is fish and food-safe
- Shade cloth or geotextile fabric
- ¼ to ⅜-inch (6–10 mm) screened rock (granite)
- Container planting mix, compost, and aged manure
- 4-inch (10 cm) perforated drainpipe
- 1-inch (3 cm) PVC for overflow pipe
- 2-inch (5 cm) PVC for water input pipe
- Silicone caulking
- Finishing nails

The size of the wood, liner, shade cloth, drainpipe, and PVC depends on the size of your wicking bed. Also, the amount of screened rock and soil mix needed is based on the volume the bed can hold. Many online tutorials are available to guide you in creating your wicking bed with other materials, such as substituting wood for recycled plastic and PVC for perforated drainpipe. These alternative materials work well, too; however, here is my favorite method:

### *Steps*

1. Level the location where the bed is to go. If the soil is rocky, add a 2-inch (5 cm) layer of sand to help protect the pond liner from being punctured.
2. Build up the sides of your raised bed with the wood planks to your desired size, ranging from 3 to 10 feet (91 cm–3 m) in length and 3 to 4 feet (91 cm–1.2 m) in width. If your bed is longer than 4 feet (1.2 m), we must brace it to prevent the sides from bowing out once filled. To stabilize long beds, nail a 2 × 2-inch (5×5 cm) wood plank vertically inside the bed every 4 feet (1.2 m). The average height of a wicking bed is 24 inches (61 m), which is most effective.
3. Add the pond liner over the bottom of the bed, ensuring that the sides of the liner extend up the sides and over the top of the bed. You can temporarily clamp the pond liner to the sides while adding other bed contents. Fill with water to a depth of 4 inches (10 cm) to ensure there are no leaks in the pond liner.
4. Place the perforated drainpipe on top of the pond liner and loop it around the inside of the bed to cover most of the bottom. Wedge the ends of the drainpipe against the side of the bed to prevent soil from entering it. Alternatively, you can use a 2-inch (5 cm) PVC pipe with multiple holes drilled along the length.

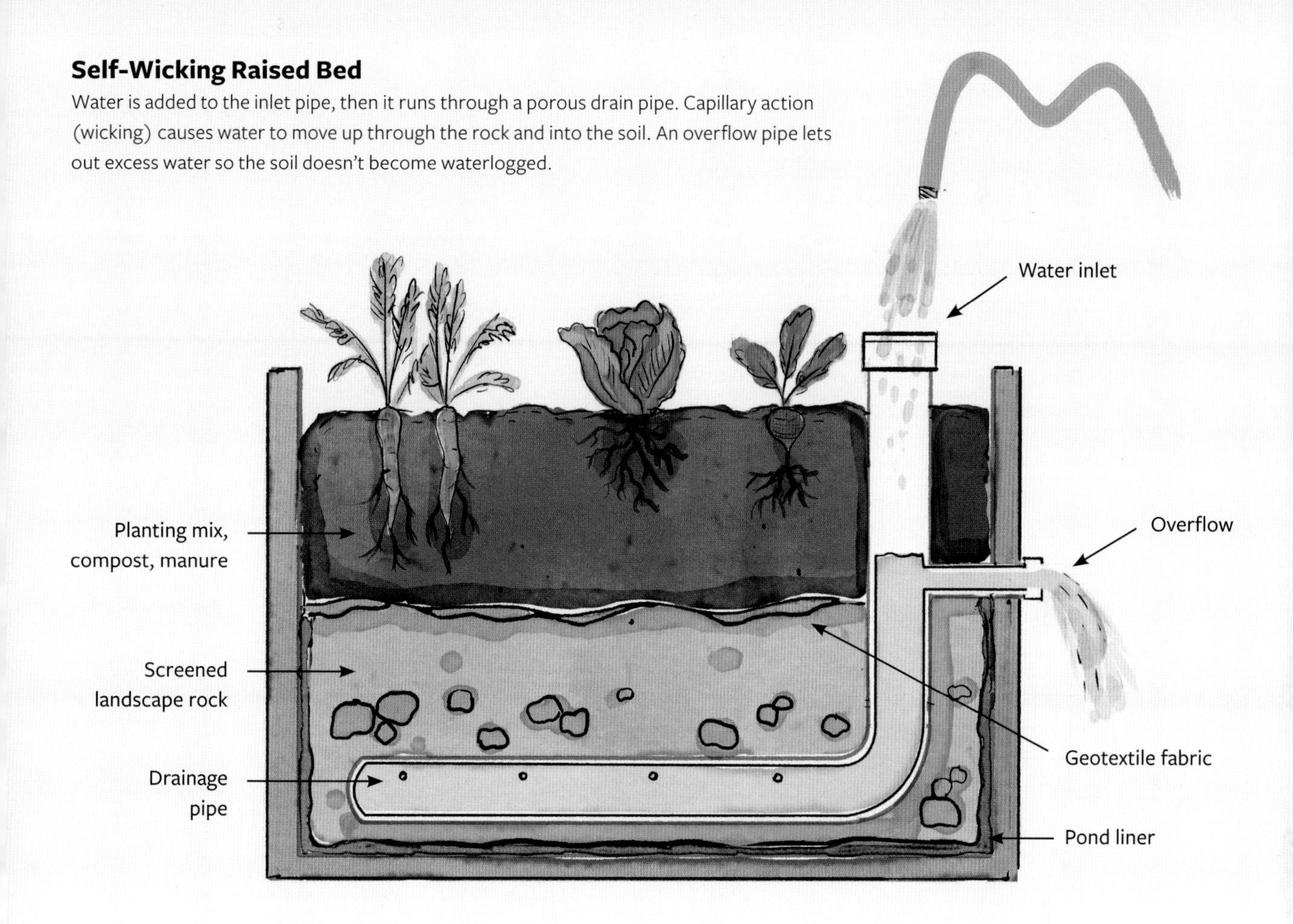


**Self-Wicking Raised Bed**

Water is added to the inlet pipe, then it runs through a porous drain pipe. Capillary action (wicking) causes water to move up through the rock and into the soil. An overflow pipe lets out excess water so the soil doesn't become waterlogged.

5. Make a 2-inch-wide (5 cm-wide) hole on top of the drainage pipe and insert the 2-inch (5 cm) PVC to extend several inches above the rim of the wicking bed. This is where the water will feed into the drainage pipe.

6. Cover the drainage pipe with ¼ to ⅜-inch (6–10 mm) screened landscape rock to an 11-inch (28 cm) depth. Screened rock ensures that the majority of the rock is of a similar size, which is very important for the wicking capability of water to move up into the soil. Fill the bed up with water to 1 inch (3 cm) below the top of the rock level. Look for any areas where puddling occurs and rake smooth.

7. For the overflow pipe, drill a 1-inch-wide (3 cm-wide) hole at the top of the rock level, through the side of the bed, on the same side where the water inlet is. Insert the 1-inch-wide (3 cm-wide) PVC pipe into the hole so that it sticks out about 3 inches (8 cm). Seal the hole around the PVC with silicone caulking to prevent it from leaking.

8. Place shade cloth (or geotextile cloth) over the rock layer and have enough to extend up the sides to the top of the bed. This helps to keep the soil and reservoir section separate but allows water to permeate into the soil.

9. Create your soil mix outside the bed, to avoid accidentally tearing the shade cloth. The soil in the bed must have a lighter consistency than that in a traditional garden bed to allow the wicking action to occur efficiently. Mix three parts of a container or potting planting mix with two parts of compost and one part of aged steer (or chicken) manure. Mix together and add to the wicking bed to a depth of 11 inches (28 cm). You can add worm castings and blood and bone meal to improve fertility. For a 24-inch-tall (61 cm-tall) bed, the reservoir and soil levels will take up approximately 22 inches (56 cm).
10. Cut the top of the water inlet PVC pipe so that it is 2 inches (5 cm) above the soil level.
11. Using finishing nails, nail the pond liner and shade cloth to the side of the bed along the inside rim and trim the excess with scissors.

## How to Use Your Wicking Bed

Before planting, we want to get the soil thoroughly moistened. Fill up the reservoir with your hose until water begins to flow out of the overflow valve, and then stop. Allow twenty-four hours for water to seep into the soil before planting. Add vegetable transplants or seeds. You will need to provide supplemental moisture for the first few weeks as the top few inches of soil are too dry to support shallow root systems until the roots grow deeper. Apply mulch around your vegetables unless the soil tends to be overly wet. Don't use tomato cages in your wicking bed, to prevent the stakes from perforating the geotextile fabric.

***To determine when it's time to refill the reservoir***

1. Stick your finger 2 to 3 inches (5–8 cm) down into the soil—if it is almost dry, it's time to fill the reservoir with water.
2. Fill the inlet pipe until you see water coming out of the overflow pipe.
3. Be careful to avoid overfilling—watch the overflow pipe as you water and stop as soon as it begins to spill water.

The frequency of adding water to your wicking bed is dependent on the weather and seasons. During rainy weather, you may need to refill your wicking bed less frequently.

Avoid adding too much organic amendments to the soil, which can become stagnant and emit odors due to the proximity to the rock/water layer. Instead, use at least 50 percent of a light container potting mix as the bulk of your soil mix and add smaller amounts of organic amendments such as compost and aged manure. Apply fertilizer at a lower rate than recommended on the bag because fertilizer isn't flushed away as it is in other growing situations.

For areas with hard water, mineral deposits can build up over time, decreasing the efficiency of your wicking bed. Mineral buildup also can occur if plants are overfertilized, as the excess salts from the fertilizer can build up. To help prevent salt and minerals from accumulating:

1. Flush out the entire wicking bed twice annually.
2. To do this, fill the inlet pipe with water until water streams out the overflow pipe.
3. Repeat daily for four to five days.

A good time to flush your wicking bed is before planting warm- and cool-season crops.

Self-wicking vegetable garden beds in Australia, made from plastic bins, so they don't require a pond liner at the base.

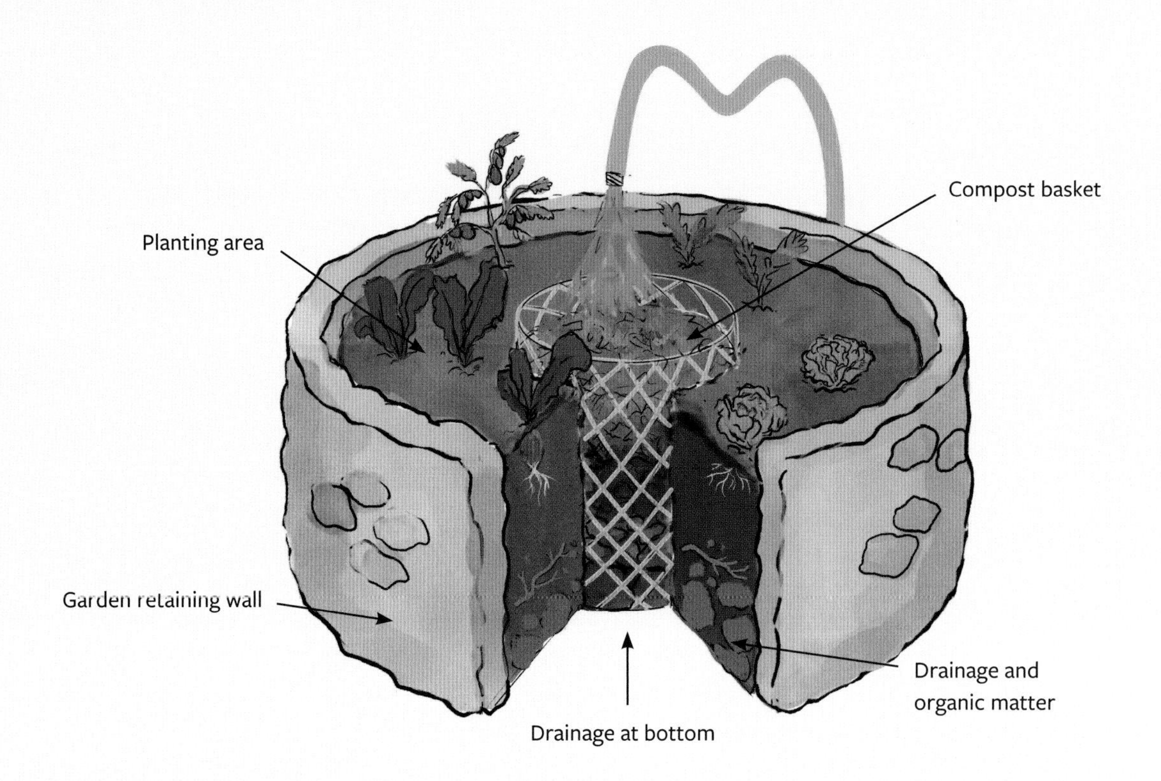


## Keyhole Gardens

A relatively new method for conserving water for growing vegetables is keyhole gardens. This method has its roots in the 1990s, when it was created to grow vegetables and herbs in drought-stricken regions in Africa that suffer from poor soil quality. Since that time, the popularity of keyhole gardens has grown throughout other regions. These gardens are mounded, circular raised beds with a notch cut out and resemble a keyhole in their shape. Their height makes gardening more accessible without bending over, and water is delivered to the center of the bed. A cylindrical "basket" lined with chicken wire or other wire mesh stands at the garden's center. Kitchen scraps and other compostable materials are added to the basket, which breaks down, adding nutrients to the surrounding soil and creating compost you can use elsewhere in the garden. Keyhole gardens allow you to grow delicious produce using less water and little to no supplemental fertilizer.

Constructing your keyhole garden isn't complicated. It begins with selecting a level location that receives at least six hours of sunlight. Avoid placing it near trees to keep unwelcome roots at bay. Be sure you have enough room to create a circular garden 6 to 8 feet (1.8 2.4 m) in width.

### How to Make a Keyhole Garden

***Materials***

- 3- to-4-foot (91–122 cm) length of string (or tape measure) to measure your circle for the garden
- Material such as stacked stone, concrete blocks, brick, or wood to build retaining walls for a keyhole garden
- Drainage materials for the base, such as cut branches or gravel
- 6- to 8-inch (15–20 cm) stone for the base of compost basket

- Chicken wire or other wire mesh to make the compost basket
- 4 to 6 metal stakes 4 feet (1.2 m) in height
- Soil mix for vegetables (topsoil, compost, aged manure)
- Mulch

### *Steps*

1. Draw out the outline of your keyhole garden. To do this, have one person stand in the middle of where you want it to go, holding the length of the string. Next, have a second person hold onto the other end of the string and lightly scratch (or mark) the soil as they rotate around the person in the center, which will help you get a circle 6 to 8 feet (1.8–2.4 m) in width.
2. Mark a notch 2 feet (61 cm) in width on the outside of the circle that tapers in width and extends to the middle—it should resemble a keyhole.
3. Add a layer of larger stone at the center of the circle to form the base of the compost basket.
4. Create a cylinder from chicken wire approximately 4 feet tall and 2.5 feet wide (1.2 m × 76 cm) to form the compost basket. Add the metal stakes around the basket to keep it sturdy and from collapsing from the weight of the soil.
5. Add drainage materials such as cut branches, twigs, or cardboard for the base of the garden, except for the notch area.
6. Build up the garden walls using your chosen material to a height of 3 feet (91 cm), except for the notch area. The compost basket should be taller than the garden to allow water to flow down through the garden.
7. Create your soil mix by adding three parts topsoil, two parts compost, and one part aged manure. Incorporate worm castings if desired to increase fertility. Then, add the soil mixture inside the walls. Be sure to keep the soil mix away from entering the basket. Mound the soil slightly above the raised walls so it reaches the 4-foot (1.2 m) height of the basket, which will help water and nutrients flow from the basket to the plants. The soil level on the outer edge of the garden should be 2 inches (5 cm) shorter than the retaining wall to prevent soil from washing over.
8. Plant your vegetables and herbs. Use smaller vegetables rather than large ones, which will take up a lot of room. Vegetables with smaller root systems, such as root vegetables, tomatoes, and leafy greens, are ideal for keyhole gardens. Provide extra water for young plants until your basket becomes full and water can be channeled throughout the garden.
9. Add a 3-inch (8 cm) layer of organic mulch to preserve soil moisture.

## Caring for Your Keyhole Garden

Add kitchen scraps, nonflowering weeds, leaves, etc., to the basket, as you would to a compost bin. As the compost breaks down, nutrients will reach the surrounding vegetables. Add new compostable material as the compost level drops down. Apply water to the basket regularly once the soil begins to dry. The keyhole garden's mounded shape helps water flow into the surrounding soil via gravity, bringing nutrients from the compost with it.

Both wicking beds and keyhole gardens can help you use less water while enjoying growing your own vegetables.

# Resources

### Shrinking Water Supply

nytimes.com/interactive/2023/08/28/climate/groundwater-drying-climate-change.html

epa.gov/watersense/outdoors

### Worldwide Water Risk Atlas

wri.org/applications/aqueduct/water-risk-atlas

### Drought Monitor

drought.gov

### Weather—Average Temperature and Precipitation

wunderground.com

### Plant Selection

gardenrant.com/2023/08/a-familiar-voice-speaks-out-on-natives-and-pollinators.html

audubon.org/native-plants

powo.science.kew.org

plants.usda.gov/home/stateSearch

### How Water Moves Through the Landscape

cmg.extension.colostate.edu/Gardennotes/262.pdf

### Drought-Resistant Soil

agwaterstewards.org/wp-content/uploads/2016/08/Drought_Resistant_Soils_ATTRA.pdf

### Soil Texture Test

agric.wa.gov.au/soil-constraints/soil-texture-estimating-hand

nrcs.usda.gov/sites/default/files/2022-11/texture-by-feel.pdf

### Gypsum vs. Compost for Soil Drainage

s3.wp.wsu.edu/uploads/sites/403/2015/03/gypsum.pdf

### Rainwater Harvesting

harvestingrainwater.com

greenbeanconnection.wordpress.com/2021/03/09/curb-cuts-for-water-harvesting-cleansing/

### Rain Barrel Vegetable Garden Irrigation Guidelines

plantaddicts.com/using-rainwater-on-vegetables-and-herbs/

### Rain Garden

epa.gov/soakuptherain/soak-rain-rain-gardens

ncei.noaa.gov/data/oceans/coris/library/NOAA/CRCP/other/other_crcp_publications/Watershed_USVI/stx_ee_hope_carton_road/130123_AttachmentC_11103.pdf

sunset.com/home-garden/landscaping/rain-garden-design

**Greywater**
harvestingrainwater.com/resource/greywater-compatible-soap-detergent-info/#choosing-greywater-compatible-soaps-detergents

**Waterwise Irrigation Methods**
wateruseitwisely.com/saving-water-outdoors/efficient-irrigation/

**Drip Irrigation Setup Detailed**
amwua.org/landscaping-with-style/design/design-the-irrigation-system

**History of Lawns in America**
nytimes.com/2019/08/09/video/lawn-grass-environment-history.html

**Calculating Water Costs for Lawn**
todayshomeowner.com/lawn-garden/guides/calculating-lawn-irrigation-costs

**Lawn Care: Recycling Clippings for Water Savings**
turf.umn.edu/news/benefits-recycling-lawn-clippings-brief-summary

**Xeriscape Conversion**
wateruseitwisely.com/blog/trade-your-water-thirsty-grass-for-a-xeriscape/

**Reconsider the Lawn**
news.climate.columbia.edu/2010/06/04/the-problem-of-lawns/

**Lawn Removal: Solarization**
bewaterwise.com/assets/190530-solarization.pdf

**Artificial Turf Considerations**
wateruseitwisely.com/saving-water-outdoors/grass-artificial-turf/10-reasons-why-artificial-turf-may-not-be-what-youre-looking-for/

**Ollas**
wateruseitwisely.com/blog/olla-irrigation/

**Ancient Low-Water Irrigation Methods**
missiongarden.org/timeline

**Self Wicking Bed**
theguardian.com/lifeandstyle/2021/nov/28/how-to-make-a-wicking-bed-a-veggie-patch-watering-solution-for-hot-australian-summers

**Keyhole Vegetable Garden**
bobvila.com/articles/keyhole-gardening/

learn.eartheasy.com/guides/how-to-build-a-raised-bed-keyhole-garden/

# Photography Credits

*All photos by author, except for the following:*

Carianne Campbell (strategichabitats.com), pages 167 (right), 168

City of Mesa Water Conservation Office (mesaaz.gov/residents/water/water-conservation), page 161 (both)

Shawna Coronado (shawnacoronado.com), pages 31 (bottom right), 86 (top), 88 (bottom), 121 (top right)

Erica Grivas, pages 15 (top), 22, 70 (bottom), 104 (top)

Grace Hensley (fashionplants.com), pages 111 (top), 136

Jennifer Humphrey, page 145

Kenny Humphrey, pages 144, 173

Todd Johnson, pages 46 (bottom), 91 (top), 101 (top), 114, 164 (right)

Byddi Lee (byddilee.com), page 160 (bottom left)

Susan Mulvihill (susansinthegarden.com), cover [vegetable garden]

Jennifer Stocker (wwwrockrose.blogspot.com), cover [courtyard landscape], pages 2, 11

Esther Smith, page 10

Grace Stufkosky (gracestufkosky.com), page 185

Andrea Whitely (andreawhitely.com), pages 17, 111 (bottom)

Robert Yontz, pages 52 (top), 60 (right), 70 (top), 151 (bottom)

Shutterstock, pages 58 (bottom), 97 (top), 143, 151, 159, 160 (bottom right)

# Location Credits

Jennifer Stocker Residence, Austin, Texas, cover photo (courtyard), pages 2, 11, 65, 74, 132

Steve and Pauline Danielson Residence, Maplewood, Minnesota, cover photo (top left), page 67

Denver Botanic Gardens Chatfield Farms (botanicgardens.org/chatfield-farms), cover photo (bottom right), page 18

Bellevue Botanical Garden (bellevuebotanical.org), page 15 (top)

The Desert Botanical Garden (dbg.org), pages 25 (top), 42, 49 (bottom), 55, 108

Ryan Residence, Phoenix, page 50 (bottom)

The Getty Center, Los Angeles, California (getty.edu), pages 52 (bottom), 54 (right), 56

Cliff Douglas Garden, Mesa, Arizona, page 53

Tane Residence, St. Paul, Minnesota, pages 72, 79, 134 (left)

Heidi Heiland Residence, Plymouth, Minnesota (growhausmn.com), pages 81 (top), 134 (right)

Gutierrez Residence, Design by Kaylee Colter at Eighty-Eight Acres (eightyeightacres.com), installation by Clint Culberson at Galaxy Gardens (galaxygardensaz.com), page 83

Founders Garden, Bisbee, Arizona, page 89

Heger Residence, Waconia, Minnesota, pages 123 (bottom right), 172

The Mission Garden, Tucson, Arizona (missiongarden.org), page 150

Rob Proctor Garden, Denver, Colorado, page 151 (top right)

Tohono Chul Gardens, Tucson, Arizona (tohonochul.org), page 151 (bottom)

# Acknowledgments

This book journey is a true compilation of many generous people who shared their knowledge and photos to help show you that being intentional with water use in the garden has many rewards without sacrificing beauty.

First, I'd like to give a big thank-you to my editor, Jessica Walliser, who asked me to write this book. As I embarked on writing, she has always been there to answer questions and give me helpful advice. The creation of a book isn't limited to its author and editor—I am grateful for the team at Cool Springs Press who took my photographs and words and turned them into a beautiful book.

I am blessed to have many garden-expert friends who willingly lent their expertise in several aspects of this book. Thank you to Shawna Coronado, Sylvia Gordon, and Erica Grivas who made valuable contributions to the drought-tolerant plant list. Because this book isn't focused on a single growing region (unlike my last one), I wanted to make sure that the photographs were representative of many different climates. I am thankful for those who graciously allowed me to use their photos—especially Susan Mulvihill and Jenny Stocker, whose photos are part of the cover. To Shawna Coronado, who would look for exam ples of wasteful irrigation to take photos for this book—I am so grateful. And a special thank-you to my dear friend, Andrea Whitely, who made sure I had photos of beautiful Australian landscapes she designed.

One of the cool parts of my career is I get to visit gardens in many places, which was useful when it came to writing this book. To those of you who allowed me to use pictures from your gardens, thank you! I would also like to extend thanks to the Desert Botanical Gardens, the Denver Botanic Gardens, the Getty Center, The Mission Garden, and Tohono Chul Gardens for permission to share the beauty of your waterwise gardens.

To my clients, students, and those who follow me on social media and my website, you serve to inspire my passion to help you create a helpful guide to water-smart gardening. Gardens should bring you joy without wasting water.

Gardening is a journey that is never finished. In my thirty-year career in the garden, plants still do things that surprise me and I enjoy the learning process. I've inadvertently killed countless plants, but so have all gardeners—each dead plant is an opportunity to learn and become a better gardener. With shrinking water resources, I am looking forward to introducing more ways to utilize water more efficiently in my own garden.

I am eternally grateful for my family who have supported me throughout the process of writing this book. From my mother, who took me on many road trips throughout the US and patiently waited for me on benches while I walked through gardens during our travels, taking photos, many of which are in this book. To my eldest daughter, Brittney, who has fully embraced the joy of gardening in Michigan and knew just the photo I needed of a rain garden. Rachele, my second-eldest daughter, is learning how to care for her garden filled with a lemon tree and flowering shrubs. And my three youngest kids, who don't have gardens of their own, yet. However, they were supportive throughout the writing process, asking me, "Mom, how is the book going?"

A special thank-you to my sister, Jennifer, who works as my assistant and manages my schedule. She has allowed me to focus on writing these past months and taken care of all other parts of my business.

Finally, to my wonderful husband, who has been my biggest cheerleader and gives me encouragement when needed, when it seemed like I would never finish this book. My garden and career would lack interest and vibrancy without his support and expertise with a shovel.

# About the Author

**Noelle Johnson** has lived in dry climates her entire life. While she was growing up in southern California, her father and grandfather introduced her to the beauty and delight of growing plants in arid regions. The desert of central Arizona has been her home for over thirty-eight years and where she learned how to garden with water conservation in mind.

Popularly known as "AZ Plant Lady," Noelle is a horticulturist and landscape consultant who lives in central Arizona with her husband. She is the author of the award-winning book, *Dry Climate Gardening*, and is a columnist for *Phoenix Home & Garden Magazine*. Noelle obtained her degree in Plant Biology, with a concentration in Urban Horticulture, from Arizona State University in 1998. Through her books, website, azplantlady.com, online classes, and speaking engagements, Noelle continues to help people learn how to create, grow, and maintain beautiful gardens that use less water.

# Index

Page numbers in *italics* indicate photographs.

## S



## T

## U



## V



## W

## X

## Y